DOING SIMP

in Your Head

DOING SIMPLE MATH in Your Head

W. J. Howard

Library of Congress Cataloging-in-Publication Data

Howard, W. J.
Doing simple math in your head / W. J. Howard
Includes index.
ISBN 1-55652-423-4

1. Arithmetic, mental. I. Title.

QA103.H6 2001 513.9
91-74030

Library of Congress Catalog Card Number: 91-74030

Cover design: Monica Baziuk

Originally published in 1992 by Coast Publishing
This edition published by Chicago Review Press, Incorporated
814 North Franklin Street
Chicago, Illinois 60610
ISBN 1-55652-423-4

Printed in the United States of America

5 4 3 2 1

To Fran, who hates math
but likes this book
(well, some of it)

CONTENTS

INTRODUCTION

If you want to multiply 17×23, chances are you'll reach for a calculator. Or you'll grab a pencil and paper. But if these things aren't handy, which is often the case, what then? Well, you can do this problem, and a variety of others that you come across daily, *in your head*—if you know how. You weren't taught how in school, where the emphasis was on pencil-and-paper techniques. This book shows you how.

School Math. In school you had to memorize addition and multiplication tables, and were exposed to rules that you may have understood only dimly. Math was something to get through and then forget about. But you can't forget about it; every day you have to cope with numbers. And one of the best means for coping—doing simple math in your head—you weren't taught in school.

School math books typically are filled with problems in which the numbers are nicely lined up, with a +, –, ×, or ÷ telling you what to do. Such problems help you become familiar with manipulating numbers—on paper. Real problems are often different. Maybe you're shopping, and wondering whether the $20 bill in your pocket will cover what you're buying. Or you're driving, and wondering how long it will be till you get home. "I have about 375 miles to go, and I'm averaging about 50 miles per hour. When will I get there?" There aren't any numbers neatly written out for you or pencil and paper to lean on.

And, as in this last example, in real life you may not *see* any numbers. You won't see 375 or 50; you'll think of them as words—"three hundred seventy-five" and "fifty"—and then do the best you can. Which may not be much if, like most people, you're used to seeing the numbers.

The Calculator. Pocket calculators are wonderful. They relieve us from the drudgery of calculation, giving us answers conveniently, quickly and quietly. However, on many occasions a calculator may not be handy. Or the problem just may not warrant a calculator; who'd want to bother with one to figure a tip in a restaurant?

Even if you have a calculator, you can't rely on it completely; it won't run itself. If you don't know how to solve a problem, you won't be sure which buttons to push. Even if you do know you may push the wrong button. You'll get the answer fast, all right—the wrong answer. But how will you *know* it's wrong, unless you have some idea of what it should be? And this you'll get by doing a little head work before you start pushing buttons.

Approximations. You're taught in school that a problem has one right answer, and that everything else is wrong. But often you don't need an exact answer; an approximation will do just fine. Many of the things we buy are priced just below a round dollar figure—$2.98, $14.95, and so on. If you think of these as $3 and $15 they're easier to handle, and you haven't lost much accuracy.

Unlike the calculator, our minds work quite naturally with approximations. If you enter a dollar figure, say $32.17, on your calculator, the 7 is treated with the same respect as the 3; to the calculator they're both numbers, and enjoy equal status. But to you, $30 is a lot more important than 7 cents, so you pay more attention to the 3 than to the 7. This natural facility helps you avoid relying entirely on a calculator. If you're about to multiply 32.17×9.63 on the calculator, first multiply 30×10 in your head to give yourself a rough idea of what the answer will be.

Using What You Know. You'd be surprised at how much you already know that you can use in figuring in your head; you're just not used to putting it together in the right way. You can divide 16 by 4: $16 \div 4 = 4$. And you can multiply 4 by 100: $4 \times 100 = 400$. Yet it probably wouldn't occur to you to use that ability to multiply 25×16. But that's all you need to do, since multiplying by 25 is the same as multiplying by 100 and dividing by 4. That is, you can multiply 25×16 by doing the two simple steps above to get the answer, 400. (If you have trouble following this, don't fret; it will be explained in more detail later.) Another example: you're thinking of buying 5 pounds of fish at $1.68 per pound. You know that 10 pounds would be $16.80, so just divide that by 2 to get the answer, $8.40.

These examples illustrate a principle regarding working problems in your head: Don't just confront the problem as it is; mold it to fit your own capabilities. In baseball and tennis you're told to "play the ball; don't let the ball play you." The same thing applies here. You want to be in charge, and tell the numbers what to do, not the other way around.

In this book, you'll find out how to put to use what you already know, and in the process learn a few new things. And you won't see many technical terms or need to memorize a long list of rules (one book has a lot of rules with names like "How to Multiply Two Two-Digit Numbers When Their First Digits Add to 10 and Their End Digits Are the Same"). What you need are a few principles—guidelines and suggestions for smoothly extending what you do now. Before you use a technique it should be so obvious to you that you can reproduce it yourself if you happen to forget.

What Is Difficult? Being able to do problems in your head is basically being able to simplify. When you can do that, the rest is easy. An illiterate man was asked how much he'd have if he worked for six hours at 35 cents an hour. He responded immediately $2.10. How did he know this? He pictured six quarters and six dimes. Four of the quarters made one dollar. Mentally, he put five of the dimes with the other two quarters to make another dollar, and had a dime left over.

You can't do everything in your head. You'd use a calculator or pencil and paper for something like 238×156. You're not going to reach the level of Johann Zacharias Dase, a nineteenth century prodigy who, among other feats, multiplied $79,532,853 \times 93,758,479$ in his head in 54 seconds! With a little practice, you'll be able to tell which problems can be easily simplified and which ones can't. You'll see, for example, that 37×46 is rather difficult, but 38×42 is easy with the right approach. The purpose of this book is to help you find the right approaches.

Contents of Book. The first chapter, "Making Things Easier," is the heart of the matter. Here you'll find some simple guidelines for making problems easier to work with—guidelines rather than rigid rules because there's often more than one way to handle a problem. Also, each problem you may come across will be a little different than the others; if you understand the principles of simplifying, you'll be better able to approach it with an open mind, rather than trying to fit it into some preconceived, memorized mold. In other words, you'll use judgment.

Sample problems and suggested solutions are in Chapter 2, where you'll have a chance to try out the ideas of Chapter 1. Some problems may not seem to apply to you, but try them anyway. It will take only a few seconds, and you may pick up something that helps in handling others. They range from simple to difficult, but not too difficult to do in your head.

You may be surprised to find the problems in "word-only" form. Many problems you face are like this. If you can solve these, you'll be able to solve the same problems more easily when you can see the numbers (the converse is not true). But if the numbers-as-words format bothers you, refer to the left side of the page, where the same problems are stated in conventional form.

The "solutions" for the problems are really suggestions, and are an important part of this book.

There's little doubt about your being able to *solve* the problems—using pencil and paper if necessary. The question is, can you do them quickly in your head? For that you need practice and suggestions on how to proceed, and that's what Chapter 2 provides.

Doing things in your head is a good test of whether you really understand numbers. In simplifying a problem, you'll be manipulating numbers quite a bit, and you want to do so quickly and easily. So Chapter 3 provides a brief "refresher course" on basic arithmetic. Refer to this chapter (and the Glossary) if you find yourself struggling. Even if you're not, some of the ideas may give you a new, helpful slant on numbers. If your math feels rather rusty now, you might start with this chapter.

Skim—or Skip—If You Like. The book is designed for easy skimming, with a running headline indicating the main idea on each page. You can flip through quickly; if something is not clear, read the page. The intent behind this layout is to avoid the twin problems of boring some while not explaining something sufficiently for others—problems that easily arise from readers' differing math backgrounds.

The fact that you're reading this book means that you can probably benefit from some of the ideas, especially those in Chapter 1. But if you think not, skip directly to the problems in Chapter 2. If you can solve them without difficulty, pass the book on to someone who needs it more than you do.

1. MAKING THINGS EASIER

You can simplify problems in various ways:...

Numbers are not rigid, static things; they're malleable and fluid, and can be shaped to meet your needs. A problem comes up involving the number 15. You don't have to work with it as such. If it suits you, you can think of it as 10 + 5, or 5 × 3, or 30/2—whatever fits your need. How do you know what will fit? You don't, without a little mental experimenting. And in experimenting a guiding light is the magic number 10, as you'll see.

A wide variety of problems will succumb to just a few easy-to-use methods: reordering numbers, rearranging them, breaking them up, using equivalents and identities, and approximating and rounding off. Always, the focus is on the number 10; it and its multiples are the easiest numbers to handle. As you review these methods, don't try to memorize; understand the logic...and then practice. When ready, turn to Chapter 2 for practice, or to Chapter 3 for help in understanding.

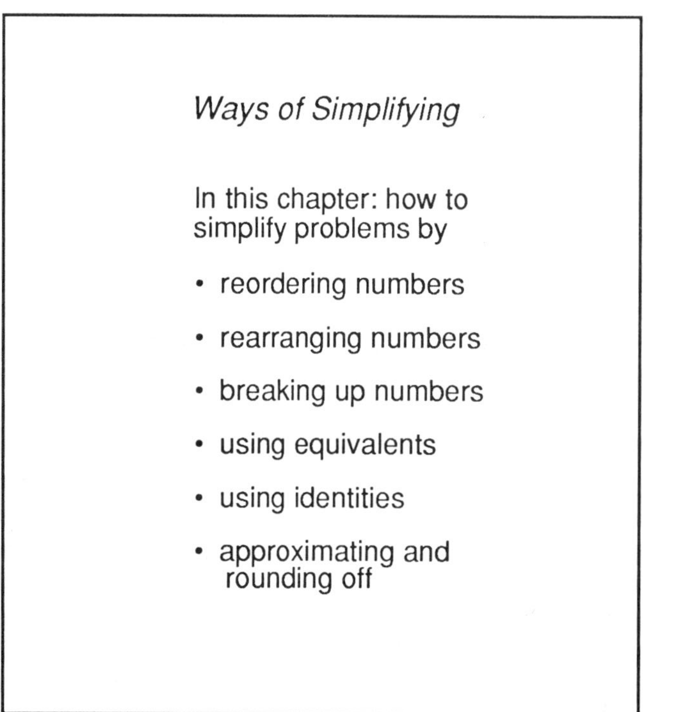

Ways of Simplifying

In this chapter: how to simplify problems by

- reordering numbers
- rearranging numbers
- breaking up numbers
- using equivalents
- using identities
- approximating and rounding off

by reordering numbers;...

Numbers come at you helter skelter. You want to add 14, 39 and 6. (Perhaps you're in a store and thinking of buying some items. In this chapter we won't worry about *why* you want to add, multiply, etc.—there will be many examples in the problems later.) The first thing to do is to reorder them to (14 + 6) + 39. Now 14 + 6 gives you 20—a multiple of 10—and 20 is easily added to 39. Result, 59—which you can do in your head in much less time than it takes to tell about it.

You have three numbers to multiply: 25 × 33 × 8. Don't try to multiply 25 × 33 (this in itself is not too difficult, but you'd still have to multiply by 8); first multiply 8 × 25, getting 200 (aha!, a multiple of 10), and then multiply this by 33. Answer: 6,600.

Maybe a problem involves a fraction: (3/4) × 48. You need to multiply 48 by 3 and divide by 4. But by all means, divide first!

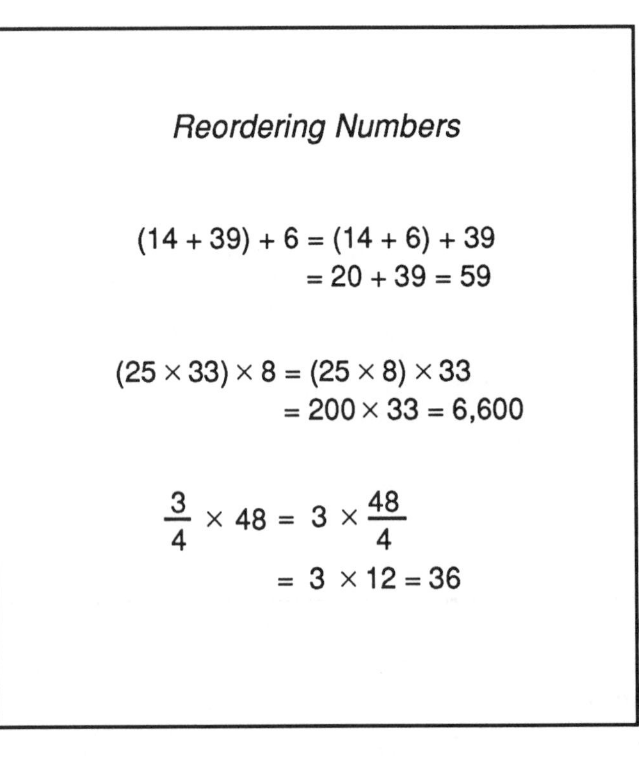

by adding and subtracting a number;...

You want to add 87 and 36. Notice that 87 is close to 90—a multiple of 10. Change 87 to 90 by adding 3, and then, to compensate, subtract that same 3 from 36, getting 33. Rearranging a problem by adding and subtracting the same number doesn't change the answer. So the original problem of 87 + 36 is now 90 + 33, which is easier. You can make it even easier by adding and subtracting 10: adding 10 to 90 gives 100—a beautiful multiple of 10—and subtracting 10 from 33 gives 23. So the original problem has now been changed to 100 + 23. It doesn't look anything like the original problem—in fact it's now no problem at all—but it will give the same answer.

In subtraction, try to change the number being subtracted to a multiple of 10. To subtract 19 from 44, add 1 to 19, giving 20, and *add* that same number (1) to 44, giving 45. The rearranged problem is now easy: 45 – 20, or 25.

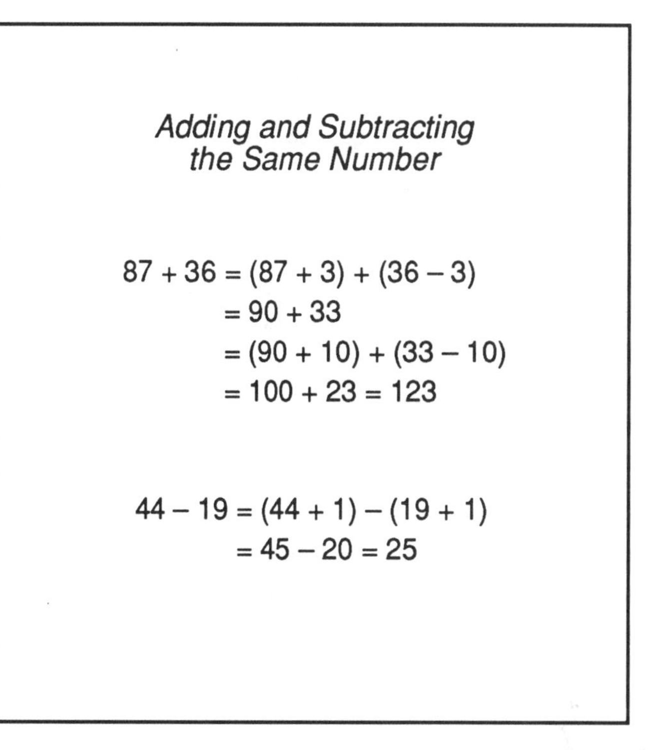

Adding and Subtracting the Same Number

$$\begin{aligned}87 + 36 &= (87 + 3) + (36 - 3)\\ &= 90 + 33\\ &= (90 + 10) + (33 - 10)\\ &= 100 + 23 = 123\end{aligned}$$

$$\begin{aligned}44 - 19 &= (44 + 1) - (19 + 1)\\ &= 45 - 20 = 25\end{aligned}$$

by multiplying and dividing by a number;...

Multiplying or dividing by 2 is easy to do in your head—a fact that has many benefits. To multiply by 5, think of multiplying by 10 and taking half of that. This gives the same result and is often easier. Suppose the problem is 5 × 14.6. Multiplying and dividing by 2 changes it to (2 × 5) × (14.6/2), so we're now multiplying by 10 (that is, 2 × 5) and dividing by 2. The result is 146/2, or 73. This also works for numbers ending in 5; multiply by 2 to get a multiple of 10, and divide by 2 to compensate. For example, 15 × 26 is changed to (2 × 15) × (26/2), or 30 × 13, or 390. We divided by 2 before multiplying here, whereas above we did so after—do whatever makes the problem easier.

When you see 2.5 or 25, think of multiplying and dividing by 4, so you'll be multiplying by 10 or 100 (and dividing by 4). A forbidding-looking 2.5 × 68 is changed to (4 × 2.5) × 68/4, or 10 × 17, or 170. (Also see fraction/decimal equivalents, later.)

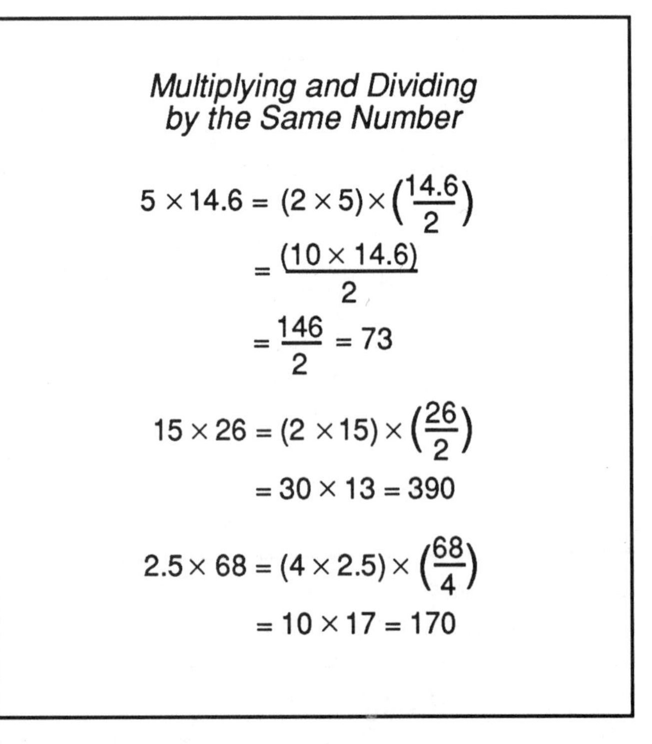

Multiplying and Dividing by the Same Number

$$5 \times 14.6 = (2 \times 5) \times \left(\frac{14.6}{2}\right) = \frac{(10 \times 14.6)}{2} = \frac{146}{2} = 73$$

$$15 \times 26 = (2 \times 15) \times \left(\frac{26}{2}\right) = 30 \times 13 = 390$$

$$2.5 \times 68 = (4 \times 2.5) \times \left(\frac{68}{4}\right) = 10 \times 17 = 170$$

by breaking up numbers...

Many problems give you a choice of how to proceed. Don't waste time waffling, or you'll defeat the purpose of making things simple. Consider 12×19. Looking for a 10 multiple, you think of 12 as $10 + 2$, so the problem is now $(10 + 2) \times 19$, which is $(10 \times 19) + (2 \times 19)$, or $190 + 38$, or 228. (If you have a question about this, see p. 95.) Alternatively, for 12×19 you might have thought of 19 as $20 - 1$, since 20 is a multiple of 10. The problem then becomes $12 \times (20 - 1)$, or $(12 \times 20) - (12 \times 1)$, or $240 - 12$, or 228 again. If you have trouble with $240 - 12$, remember to change the problem so you're subtracting a multiple of 10. In this case, subtract 2 from each number: $240 - 12 = (240 - 2) - (12 - 2)$, or $238 - 10$, which is 228.

On the previous page we multiplied 15×26 by multiplying by 2 and dividing by 2. Instead, we might have broken up 15 into $10 + 5$; the problem becomes $(10 + 5) \times 26$, which is $260 + 130$, or 390.

Breaking Up Numbers

$$
\begin{aligned}
12 \times 19 &= (10 + 2) \times 19 \\
&= (10 \times 19) + (2 \times 19) \\
&= 190 + 38 = 228
\end{aligned}
$$

or

$$
\begin{aligned}
12 \times 19 &= 12 \times (20 - 1) \\
&= (12 \times 20) - (12 \times 1) \\
&= 240 - 12 = 228
\end{aligned}
$$

$$
\begin{aligned}
15 \times 26 &= (10 + 5) \times 26 \\
&= (10 \times 26) + (5 \times 26) \\
&= 260 + 130 = 390
\end{aligned}
$$

in various ways,...

The numbers 9 and 11 are easy to handle, since they're both so close to 10; think of 9 as $10-1$ and 11 as $10+1$. To multiply by 9, first multiply the number by 10 and then subtract the number: $9 \times 26 = (10-1) \times 26 = (10 \times 26) - (1 \times 26)$, which is $260 - 26$. Some may wish to simplify even further by adding 4 to each number: $260 - 26 = (260+4) - (26+4)$, which is $264 - 30$, or 234.

To multiply by 11, multiply by 10 and add the number: $11 \times 26 = (10+1) \times 26$, which is $(10 \times 26) + (1 \times 26)$, or $260 + 26$, or 286.

You can even handle larger numbers, like 99 and 101, since they're so close to 100. For example, $99 \times 84 = (100-1) \times 84$, or $(100 \times 84) - (1 \times 84)$, or $8{,}400 - 84$. You can simplify to $(8{,}400 + 16) - (84 + 16)$, or $8{,}416 - 100$, or 8,316. If the problem is 101×84 it's even easier: $101 \times 84 = (100+1) \times 84$, or $(100 \times 84) + (1 \times 84)$, or $8{,}400 + 84$, or 8,484.

Breaking Up Numbers (cont'd)

$$\begin{aligned} 9 \times 26 &= (10-1) \times 26 \\ &= (10 \times 26) - (1 \times 26) \\ &= 260 - 26 = 234 \end{aligned}$$

$$\begin{aligned} 11 \times 26 &= (10+1) \times 26 \\ &= (10 \times 26) + (1 \times 26) \\ &= 260 + 26 = 286 \end{aligned}$$

$$\begin{aligned} 99 \times 84 &= (100-1) \times 84 \\ &= (100 \times 84) - (1 \times 84) \\ &= 8{,}400 - 84 = 8{,}316 \end{aligned}$$

$$\begin{aligned} 101 \times 84 &= (100+1) \times 84 \\ &= (100 \times 84) + (1 \times 84) \\ &= 8{,}400 + 84 = 8{,}484 \end{aligned}$$

including factoring;...

When multiplying or dividing, it's often useful to factor numbers (composite numbers can be factored into their components: $6 = 2 \times 3$, for example—see p. 97). Suppose you want to divide 78 by 6. You can perhaps do this directly in your head, but if you have difficulty, divide first by 2 and then by 3. Dividing by 2 gives you 39, and dividing this by 3 gives you the answer, 13.

Another problem: divide 396 by 18. Since $18 = 2 \times 3 \times 3$, you can divide first by 3, getting 132, then by 2, getting 66, then by 3 again, getting the final answer, 22. You might also solve this directly by noting 20×18 is 360, and there's still 2×18 to go.

Suppose you want to multiply 28×13. Since $28 = 7 \times 2 \times 2$, you can restate this problem as $(7 \times 2 \times 2) \times 13$, which is $7 \times (2 \times 26)$, or 7×52. For 7×52 you can break up 52 into $50 + 2$ and multiply separately: $(7 \times 50) + (7 \times 2)$ is $350 + 14$, or 364.

Factoring Numbers

$$\begin{aligned} 78 \div 6 &= 78 \div (2 \times 3) \\ &= (78 \div 2) \div 3 \\ &= 39 \div 3 = 13 \end{aligned}$$

$$\begin{aligned} 396 \div 18 &= 396 \div (2 \times 3 \times 3) \\ &= (396 \div 3) \div (2 \times 3) \\ &= (132 \div 2) \div 3 \\ &= 66 \div 3 = 22 \end{aligned}$$

$$\begin{aligned} 28 \times 13 &= (7 \times 2 \times 2) \times 13 \\ &= 7 \times (2 \times 26) = 7 \times 52 \\ &= (7 \times 50) + (7 \times 2) \\ &= 350 + 14 = 364 \end{aligned}$$

by using fraction/decimal equivalents...

Fractions are surprisingly useful, particularly in multiplication problems (see p. 114). To multiply 50×38, it's easy to multiply 100×38 and take half of that—1,900. Another problem: 0.25×436. You'd have quite a time multiplying this in your head. But realizing that 0.25 is 1/4, you just divide 436 by 4 and have the answer, 109.

Once you become familiar with fractions and their decimal equivalents, you find many opportunities to use them. It's not just in the form of numbers less than one that fractions are useful. If the second problem above had been 2.5×436, you'd do the same thing—divide by 4—but then, since 2.5 is 10 times 0.25, you'd have to multiply by 10 to get the answer: 1,090. With practice, you get in the habit of looking for numbers that correspond to fractional equivalents, knowing you can multiply or divide by a power of 10 (10, 100, etc.—see p. 98) to make the answer come out right.

Using Fraction/Decimal Equivalents

$$50 \times 38 = \frac{1}{2} \times 100 \times 38$$
$$= \frac{38}{2} \times 100 = 1{,}900$$

$$0.25 \times 436 = \frac{1}{4} \times 436$$
$$= \frac{436}{4} = 109$$

$$2.5 \times 436 = \frac{1}{4} \times 10 \times 436$$
$$= \frac{436}{4} \times 10$$
$$= 109 \times 10 = 1{,}090$$

to convert awkward numbers...

Some fractional equivalents are familiar; it's common knowledge that 1/2 = 0.5, 1/4 = 0.25, and 3/4 = 0.75. Also, 1/3 = 0.33 and 2/3 = 0.67 (to two places). A problem involving these numbers can often be handled quickly. For example: 67×39 seems at first like a difficult problem. But think of 67 as $(2/3) \times 100$. Note also that 1/3 of 39 is 13, so 2/3 is 26, and you have an approximate answer right away: $(2/3) \times 100 \times 39$ is 2,600. The answer is not exact since $(2/3) \times 100$ is not exactly 67.

As for the less familiar equivalents: since you know that 1/4 is 0.25, you also know that 1/8, being half of 1/4, is 0.125. So if a problem involves 125, you'll think of dividing by 8 and multiplying by the proper power of 10. By dividing by 2 again, you also know that 1/16 is 0.0625. Or take 1/3, which you know is 0.333; 1/6 is half of that, or 0.167, and 1/12 is half of that, or 0.083. Any or all of these relations can come in handy...if you think of them.

Using Fraction/Decimal Equivalents (cont'd)

Approximating:

$$67 \times 39 = \frac{2}{3} \times 100 \times 39 \text{ (approx.)}$$
$$= \frac{39}{3} \times 2 \times 100$$
$$= 13 \times 2 \times 100 = 2{,}600$$

Finding other equivalents:

$$\frac{1}{8} = \frac{1}{2} \times \frac{1}{4} = \frac{1}{2} \times 0.25 = 0.125$$
$$\frac{1}{6} = \frac{1}{2} \times \frac{1}{3} = \frac{1}{2} \times 0.333 = 0.167$$

to simpler ones;...

You can do some difficult-looking problems in your head using fraction/decimal equivalents. How about 12.5×17? Since 1/8 is 0.125, 12.5 is $(1/8) \times 100$, and 12.5×17 is $(1/8) \times 100 \times 17$. Now 17/8 is 2 1/8, or 2.125. So 12.5×17 is 2.125×100, or 212.5. It takes longer to describe it than to do it. Notice that this works even though 8 does not evenly divide 17. Another: $56 \div 17.5$ looks impossible until you realize $17.5 = 10 \times 1.75$, and $1.75 = 7/4$. So (see p. 106) $56 \div 17.5 = 56 \times (4/7) \times (1/10)$, which is $(8 \times 4)/10$, or 3.2. Not an easy problem.

If you multiply 200×300, you'll first think "$2 \times 3 = 6$," then start wondering how many zeros to tack on. This is how to handle a problem using fraction/decimal equivalents. Think about the numbers or digits first, without worrying about size of the answer. This lets you make useful associations. Size is just a matter of factors of 10—how many zeros or where the decimal point goes.

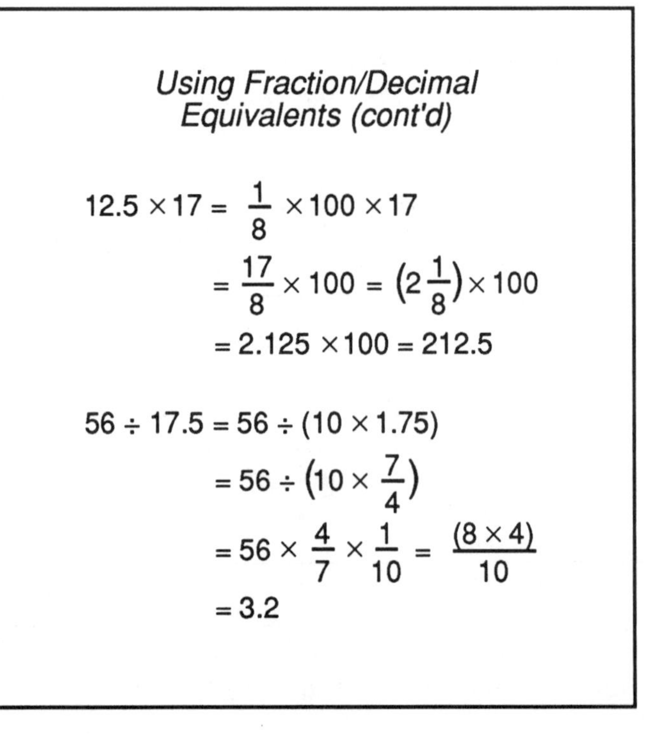

and by using identities...

Try multiplying $5 - 3$ by $5 + 3$. Yes, the answer will be 2×8, or 16, but let's multiply it out: $(5 - 3) \times (5 + 3) = (5 \times 5) + (5 \times 3) - (5 \times 3) - (3 \times 3)$. Since the middle terms cancel, this is just $(5 \times 5) - (3 \times 3)$, or $25 - 9$, or 16. To save space, it's usual to write 5×5 and 3×3 as 5^2 and 3^2, respectively. So we've found that $(5 - 3) \times (5 + 3) = 5^2 - 3^2$.

We don't need this kind of help in multiplying 2 by 8, but no matter what the two numbers are, the same relation will hold. For *any* two numbers, say a and b, $(a - b) \times (a + b) = a^2 - b^2$—as you can see by multiplying out. This relation is very useful for larger numbers. (We've sneaked in a little algebra here—using letters for numbers.)

Say you want to multiply 17×23. Looking for a multiple of 10, you see they're both near 20, so you can think of this as $(20 - 3) \times (20 + 3)$. And this is simply $20^2 - 3^2$, or $400 - 9$, or 391.

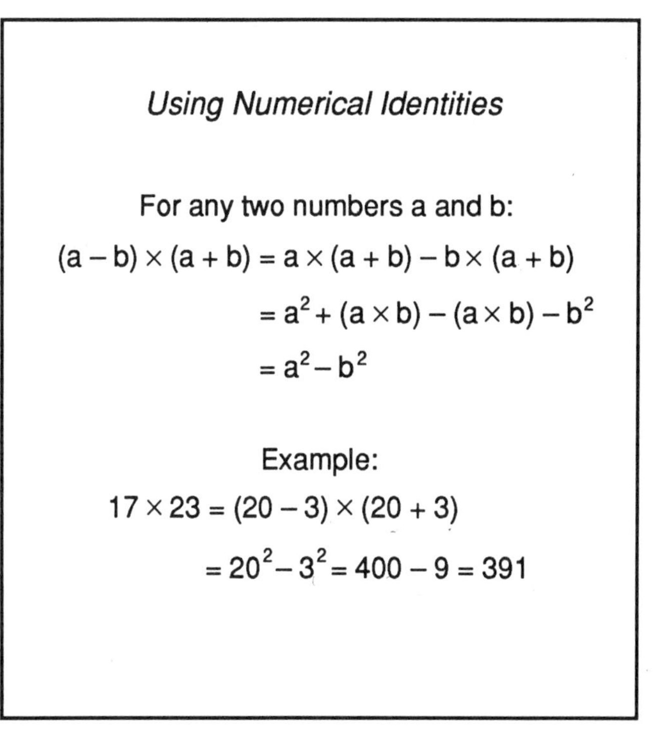

that apply to any numbers.

Flushed with success at finding one useful relation, let's look for another. Rather than $(5 - 3) \times (5 + 3)$, suppose we had $(5 + 3) \times (5 + 3)$; what then? Multiplying out, we have $(5 \times 5) + (5 \times 3) + (5 \times 3) + (3 \times 3)$. This time, the middle terms don't cancel, so we're left with $5^2 + 2 \times (5 \times 3) + 3^2$, or $25 + 30 + 9$, or 64—what we expected, since $8^2 = 64$. We've found that $(5 + 3)^2 = 5^2 + 2 \times (5 \times 3) + 3^2$.

Again, 5 and 3 are not unique—this same relation holds for any two numbers a and b: $(a + b)^2 = a^2 + 2 \times (a \times b) + b^2$—no big deal, but still useful. For example, you want to "square" 41 (that is, multiply 41×41). Think of it as $(40 + 1)^2$, which is $40^2 + 2 \times 40 \times 1 + 1^2$, or $1{,}600 + 80 + 1$, or 1,681.

What about $(a - b)^2$? Multiplying out, you see that $(a - b)^2 = a^2 - 2 \times (a \times b) + b^2$. So, for example, $28^2 = (30 - 2)^2 = 30^2 - 2 \times (30 \times 2) + 2^2$, or $900 - 120 + 4$, or 784. Another tool to add to our repertoire.

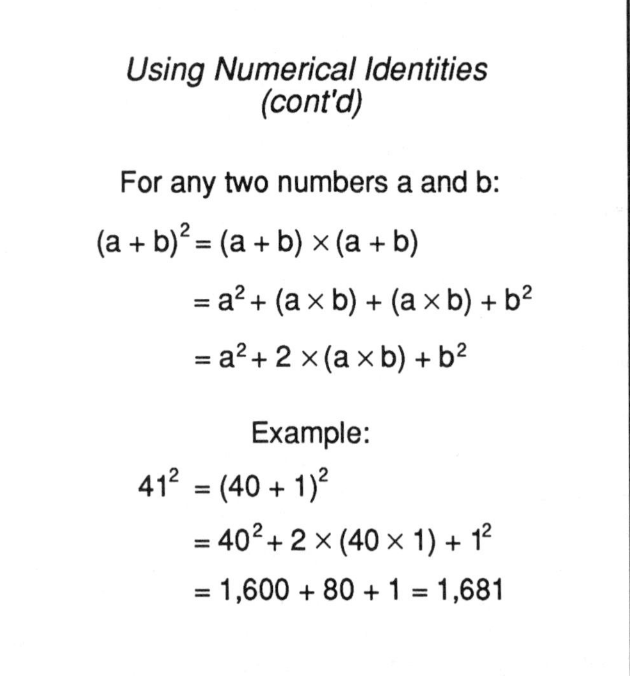

Round off and approximate...

With each digit worth ten times the digit to its right, our decimal system makes it easy to focus on the ones that count. But you don't want to forget entirely about the right-hand digits; if you go through life thinking of $1.75 as $1, you're in trouble. So what you do is "round off" (see p. 113). To round off a number, drop unwanted digits on the right. If the first digit dropped is 5 or greater, increase the preceding digit by 1; otherwise leave it unchanged. For example, to three places, 1/3 is rounded to 0.333, and to two places, 0.33; 2/3 is rounded to 0.667 and 0.67.

Other examples: $1.75, rounded to the nearest dime, is $1.80, since the digit dropped is 5; to the nearest dollar, $1.75 rounds to $2. A bill of $12.87 (next page) rounds to $12.90 to the nearest dime, $13 to the nearest dollar, and $10 to the nearest ten dollars. Rounding off just makes clear what we mean when we say $12.87 is "about $13."

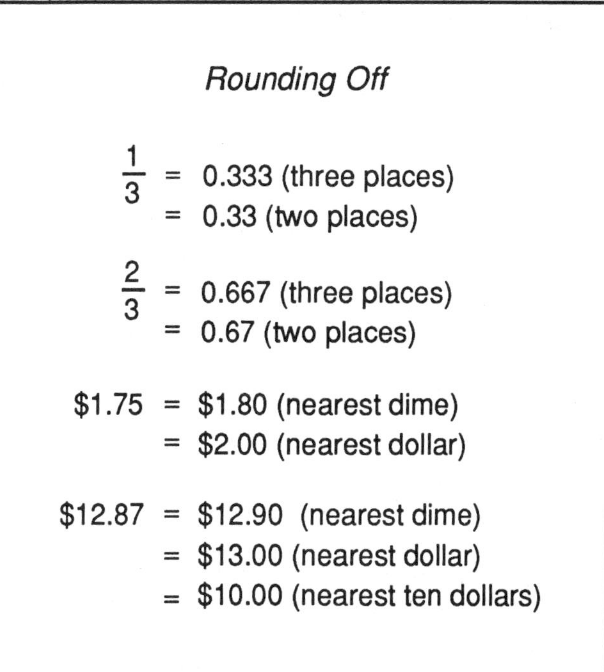

Rounding Off

$\frac{1}{3}$ = 0.333 (three places)
= 0.33 (two places)

$\frac{2}{3}$ = 0.667 (three places)
= 0.67 (two places)

$1.75 = $1.80 (nearest dime)
= $2.00 (nearest dollar)

$12.87 = $12.90 (nearest dime)
= $13.00 (nearest dollar)
= $10.00 (nearest ten dollars)

where appropriate,...

If you're figuring how much to tip in a restaurant, you certainly don't need to work it out to the last penny—unless you want to establish a reputation as a penny-pincher, even while trying to be generous with a tip. Just an approximate amount is all that's expected. Often you want to give a 15% tip. The easiest way to do this is to first figure 10% by moving the decimal point one place to the left (see pp. 110 and 116), then take half of this, which will be 5%, and add the two. Round off as you proceed. Suppose your bill is $12.87; 10% of this is about $1.30, and half of $1.30 is $0.65. Adding the two, you have $1.95. So if you leave about $2 you'll be doing what you wanted to do.

Many items are priced at just less than an even dollar figure, for sales purposes: $8.99, $13.95, and so on. You don't usually care about the pennies, so just round these up to the nearest dollar: $8.99 + $13.95 is about $9 + $14, or $23.

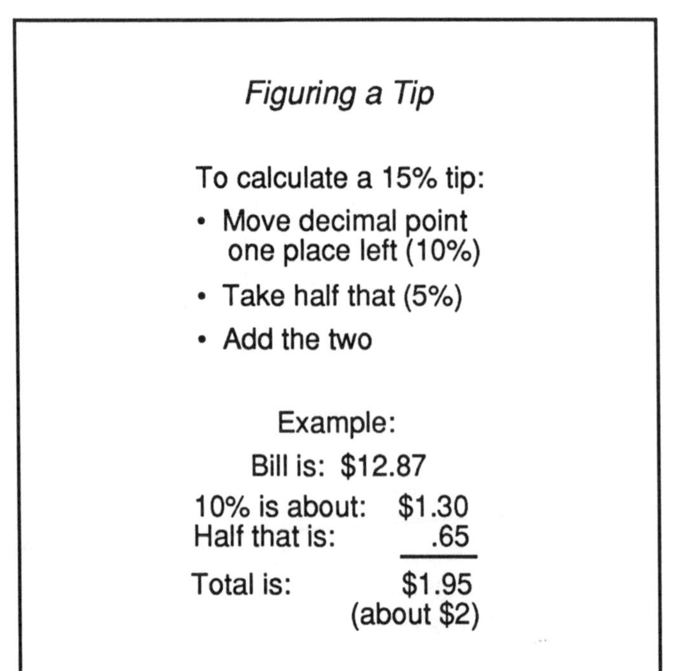

but be aware that errors can creep in.

It's easier, of course, to work with numbers when you approximate and round off—that's why you do it. But you pay a penalty, as your answers are less accurate. So do it when accuracy is not at a premium. For example, in adding 873 and 465, you might use 900 and 500, getting 1,400. The correct answer of 1,338 is quite a bit different. Does it matter? You must be the judge.

Larger errors can be made in multiplying. For 67 × 31, you might try 70 × 30, or 2,100. Not bad: the correct answer is 2,077. But suppose you simplified 72 × 26 to 70 × 30. The result, 2,100 again, is quite a bit larger than the correct answer, 1,872. There is too much change in the smaller number: 26 became 30. Instead of adding 26 72's together (p. 84), you're adding 30, or four more of the larger number than there should be. The principle: be careful in rounding off numbers before multiplying—particularly the smaller of two numbers.

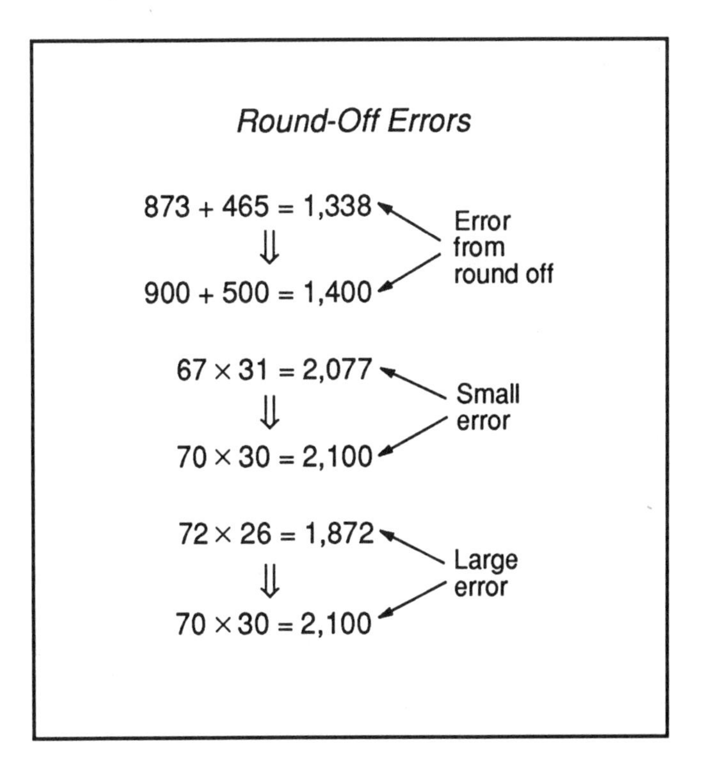

Ready for problems? Mental exercises help.

Doing math in your head is unfamiliar ground; even with the suggestions given, you may have trouble. One thing you can do to feel more comfortable is to use idle moments to practice. First, a logical extension to the "tables" you learned long ago is to add and multiply two-digit numbers. Start with easier ones like 11, 12 and 13 and work up. Try doing it in different ways, using the suggestions in this chapter and perhaps some of your own. Next, become familiar with fraction/decimal equivalents, for fractions with denominators ranging from 2 to 12. Don't try to memorize; look for relations (such as 5/12 = 3/12 + 2/12, which is 1/4 + 1/6, or 0.25 + 0.167, or 0.417). Finally, devise problems for yourself, such as converting monthly to annual bills or salaries. You can do 12 × \$750 in three ways: (10 + 2) × \$750, 12 × (\$700 + \$50), and (3/4) × 12 × \$1,000. After a while, you'll find it so easy and satisfying you'll wonder why you haven't been using your head more all along.

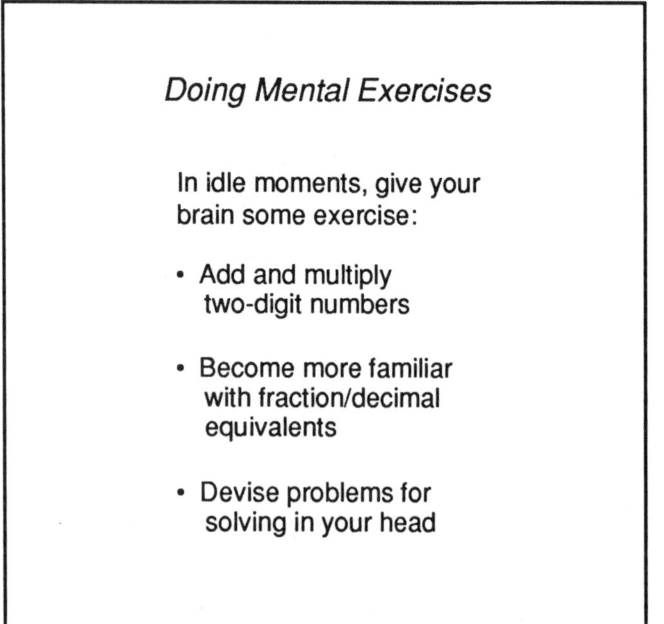

2. PROBLEMS AND SOLUTIONS

The usual "book" problems don't suit our purpose,...

Problems can be stated in different ways; shown here are three common types. You'll find many problems of Type 1 in arithmetic books. We've all done these and gained some good practice with numbers. You also find problems of Type 2, where the numbers are not lined up vertically and are a little harder to work with. And then there are "word problems"—Type 3—which try for realism.

As mentioned in the Introduction, problems that you come across in daily life are not like this. They're certainly different than Types 1 and 2, since *you* have to decide what to do—there's no +, –, × or ÷ sign to tell you. But they also differ from Type 3; real problems often don't even give you numbers, as such, to work with; you have to deal with numbers as words in your mind. Notice that with Type 3, the numbers appear as numbers, not as words. How can problems be stated so they're more like those you'll be doing in your head?

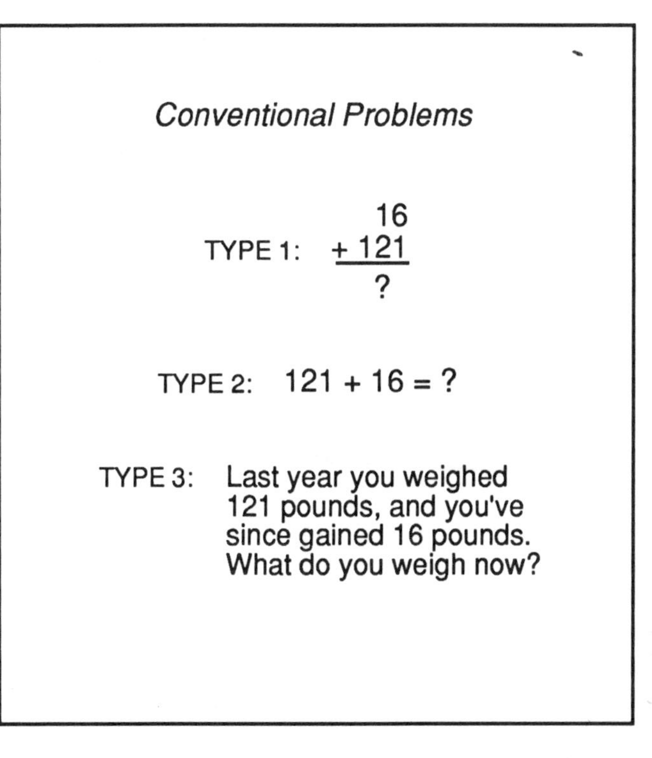

so problems are stated differently here.

Let's look again at the problem on the previous page. A friend calls you on the phone and says "I weighed one hundred twenty-one pounds last year and have since gained sixteen pounds." You might want to do this in your head, as it's not difficult. But in this problem—and in many others in real life—the numbers come at you as *words*. You have to *visualize* the numbers if you can.

To work problems in your head, you need practice with numbers as words. But you may also need an alternative to fall back on until you become used to visualizing. So the problems in this chapter appear twice: On the right of the page, in the box, the problem is entirely in words, as shown here (right). The essential facts are in tabular form so you don't have to do a lot of reading. If this form gives you trouble, refer to the left side of the page. There the same problem appears in conventional form, with numbers shown as numbers.

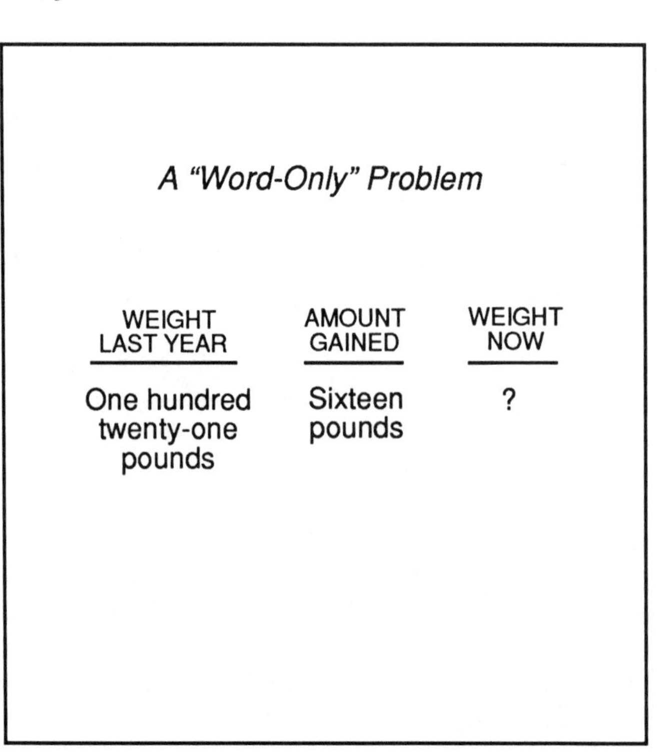

A "Word-Only" Problem

WEIGHT LAST YEAR	AMOUNT GAINED	WEIGHT NOW
One hundred twenty-one pounds	Sixteen pounds	?

Problems – 1

Let's begin with a few simple problems. Try to work them in your head by first looking only at the box at the right. If you have trouble, look at the more conventional statements below. When done, turn the page for suggestions.

You're comparing your household bills for four months.

a) In January you had a phone bill of $23 and utilities of $64. What was the total of these bills?

b) In February your phone bill was $34 and your utilities, $159. What was the total?

c) In March your phone bill was $18 and your utilities, $135. The total?

d) In April your phone bill was $57 and your utilities, $166. The total?

Household Bills

		PHONE BILL	UTILITY BILL	TOTAL
a)	Jan	Twenty-three dollars	Sixty-four dollars	?
b)	Feb	Thirty-four dollars	One hundred fifty-nine dollars	?
c)	Mar	Eighteen dollars	One hundred thirty-five dollars	?
d)	Apr	Fifty-seven dollars	One hundred sixty-six dollars	?

Solutions – 1

These first problems are simple additions, to give you a feeling for doing problems in your head.

a) This shouldn't give you much trouble, even without seeing the numbers. The total is $87.

b) This one is a little more difficult. The total would be the same if the utility bill were $160 and the phone bill $33. Adding, you get $193.

c) Increase the phone bill to $20, and decrease the utilities by the same amount ($2) to $133. The total you now get is $153.

d) Change the numbers in your head to $60 and $163. Since 16 + 6 is 22, you know that 160 + 60 is 220, so the sum of $60 and $163 is $223. (Or if you like, think of 160 + 60 as (160 + 40) + 20, which is 200 + 20, or 220.)

a)
```
 $23
+64
 $87
```

b)
```
 $ 34          $ 33
+159    ⇒    +160
   ?          $193
```

c)
```
 $ 18          $ 20
+135    ⇒    +133
   ?          $153
```

d)
```
 $ 57          $ 60
+166    ⇒    +163
   ?          $223
```

Problems – 2

You're talking to a friend about car insurance, and find that you're paying different amounts for similar features.

a) You find that for one part of your insurance, you're paying $64 and he/she is paying $47. How much more are you paying?

b) For another part, you're paying $83 and he/she is paying $69. How much more are you paying?

c) For a third part, you're paying $221 and he/she is paying $176. How much more are you paying?

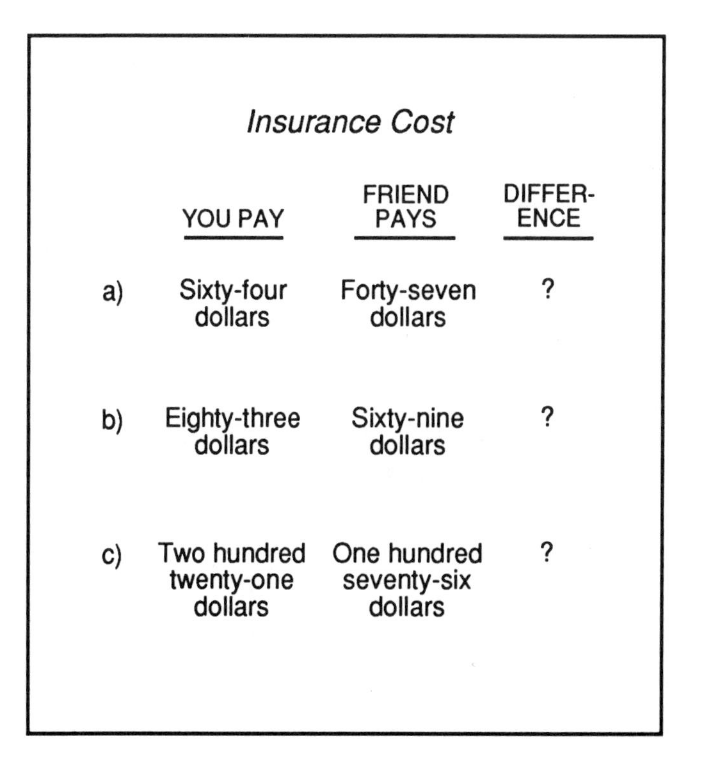

Insurance Cost

	YOU PAY	FRIEND PAYS	DIFFER-ENCE
a)	Sixty-four dollars	Forty-seven dollars	?
b)	Eighty-three dollars	Sixty-nine dollars	?
c)	Two hundred twenty-one dollars	One hundred seventy-six dollars	?

Solutions – 2

These are simple problems in subtraction. It's usually a good idea to modify a subtraction problem so that the number being subtracted is a multiple of 10, such as 10 itself, 20, 30, etc.

a) Change \$47 to \$50 and \$64 to \$67; the difference will remain the same since you have added the same number, \$3, to both amounts. Now it's easy to subtract \$50 from \$67, getting \$17.

b) Proceed in the same way: add a dollar to each amount, getting \$84 and \$70. Then subtract \$70 from \$84, getting \$14.

c) Change the two numbers to \$225 and \$180 by adding \$4 to each. Subtract \$180 from \$225, getting \$45. You can make it even easier to subtract \$180 from \$225 by adding \$20 to each, changing the problem to \$245 – \$200.

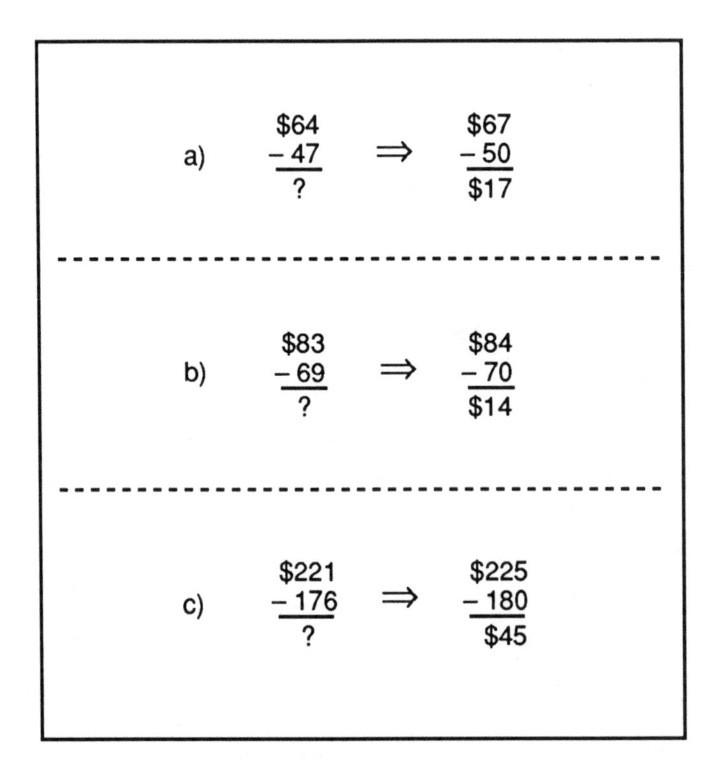

Problems – 3

You're in a drugstore, thinking of buying two items.

a) The items are priced at $6 and $17. What would be the total cost of the two items?

b) The two items are priced at $4.50 and $7.50. What would be the total cost?

c) The two items are priced at $6.95 and $1.95. What would be the approximate total cost? The exact total?

d) The two items are priced at $8.88 and $15.75. What would be the approximate total cost? The exact total?

Cost of Two Items

	ITEM 1 PRICE	ITEM 2 PRICE	TOTAL COST
a)	Six dollars	Seventeen dollars	?
b)	Four fifty	Seven fifty	?
c)	Six ninety-five	One ninety-five	Approx. ? Exact ?
d)	Eight eighty-eight	Fifteen seventy-five	Approx. ? Exact ?

Solutions – 3

More simple addition problems. These also show that you can often approximate, and, without much more effort, get an exact answer.

a) You shouldn't have any trouble adding $17 and $6 in your head: total, $23.

b) Add $4 and $7 separately, getting $11; then add $0.50 and $0.50, getting $1.00. Finally, add $11 and $1, getting $12.

c) First get the approximate total by thinking of $6.95 as $7 and $1.95 as $2; then $7 + $2 is $9. Since you've added in a nickel too much for each item, or a total of a dime, subtract a dime from the first answer. The exact amount is $8.90.

d) Approximate the two items as $9 and $16 and add, getting $25. For the exact answer, subtract $0.12 + $0.25, or $0.37, from $25, getting $24.63.

a)	$17	
	+ 6	
	$23	

b)	$4.50	⇒	$ 4	$0.50	$11
	+7.50		+ 7	+ .50	+ 1
	?		$11	$1.00	$12

			Approx.	Exact	
c)	$6.95	⇒	$7	$0.05	$9.00
	+1.95		+2	+ .05	– .10
	?		$9	$0.10	$8.90

			Approx.	Exact	
d)	$ 8.88	⇒	$ 9	$0.12	$25.00
	+15.75		+16	+ .25	– .37
	?		$25	$0.37	$24.63

Problems – 4

Here are some problems in making change in a grocery store.

a) You've so far bought canned goods totaling \$3.00 and produce totaling \$0.80. What will be your change from a \$5 bill?

b) Same situation, but your canned goods total \$16.25 and your produce, \$2.70. What will be your change from a \$20 bill?

c) Same again, with a canned goods total of \$4.89 and produce total of \$3.54. Change from a \$10 bill?

Making Change

	CANNED GOODS	PRODUCE	CHANGE
a)	Three dollars	Eighty cents	From five dollar bill?
b)	Sixteen twenty-five	Two seventy	From twenty dollar bill?
c)	Four eighty-nine	Three fifty-four	From ten dollar bill?

Solutions – 4

a) Add \$3.00 and \$0.80; result, \$3.80. To figure change from a \$5 bill, first work up to \$4 by adding \$0.20. Then add \$1 to get to \$5. Your change will be the sum of \$0.20 and \$1, or \$1.20.

b) Add \$16 and \$2, getting \$18. Then add \$0.25 and \$0.70, getting \$0.95. (Or do both these additions at once if you can.) The total is \$18.95. To figure change from a \$20 bill, add 5 cents to \$18.95 to get to \$19.00, then add \$1 to get to \$20. Your change will be \$0.05 + \$1.00, or \$1.05.

c) Think of \$4.89 as \$5, realizing you're 11 cents too high. Adding \$5 and \$3.54, you get \$8.54. Now subtract that 11 cents, getting \$8.43. Figure your change by adding 7 cents to get to \$8.50, then \$0.50 to get to \$9, then \$1 to get to \$10. Your total change will be \$1.57.

$$\text{a)}\quad \begin{array}{r} \$3.00 \\ +\ .80 \\ \hline \$3.80 \end{array} \Rightarrow \begin{array}{l} \$3.80 + \$0.20 = \$4.00 \\ \$4.00 + \underline{\$1.00} = \$5.00 \\ \qquad\quad \$1.20 \text{ (Change)} \end{array}$$

$$\text{b)}\quad \begin{array}{r} \$16.25 \\ +\ 2.70 \\ \hline \$18.95 \end{array} \Rightarrow \begin{array}{l} \$18.95 + \$0.05 = \$19.00 \\ \$19.00 + \underline{\$1.00} = \$20.00 \\ \qquad\qquad \$1.05 \text{ (Change)} \end{array}$$

$$\text{c)}\quad \begin{array}{r} \$4.89 \\ +3.54 \\ \hline ? \end{array} \Rightarrow \begin{array}{l} \$5.00 + \$3.54 = \$8.54 \\ \$8.54 - \$0.11 = \$8.43 \end{array}$$

$$\Downarrow$$

$$\begin{array}{l} \$8.43 + \$0.07 = \$8.50 \\ \$8.50 + \$0.50 = \$9.00 \\ \$9.00 + \underline{\$1.00} = \$10.00 \\ \qquad\quad \$1.57 \text{ (Change)} \end{array}$$

Problems – 5

You're in a store, thinking of buying several items, and you're wondering about the approximate cost of the three items.

a) The prices for the three items are $12, $16, and $8. What will you pay for the items? (Here, you can easily get the exact cost, so you don't need to approximate.)

b) The prices are $3.99, $24.88, and $5.95. What will be the approximate total cost of the three items?

c) The prices are $8.88, $56.75, and $12.90. What will be the approximate total cost?

d) The first item is on sale for two for $15.95, and you want one. The prices for the two other items are $69.90 and $41.85. What will you pay, approximately, for the three items?

Cost of Three Items

	ITEM 1 PRICE	ITEM 2 PRICE	ITEM 3 PRICE	TOTAL COST
a)	Twelve dollars	Sixteen dollars	Eight dollars	? (Exact)
b)	Three ninety-nine	Twenty-four eighty-eight	Five ninety-five	? (Approx.)
c)	Eight eighty-eight	Fifty-six seventy-five	Twelve ninety	? (Approx.)
d)	Two for fifteen ninety-five	Sixty-nine ninety	Forty-one eighty-five	? (Approx.)

Solutions – 5

a) To get the total price of the three items, it's best to add $12 and $8 first, as they sum to $20, a multiple of 10. Adding $16 to $20, you find the total to be $36.

b) Approximate the three prices as $4, $25, and $6. You can add these in any order, but it's easiest to add $4 and $6, getting $10. Then add $25 to get an approximate total cost of $35.

c) The three prices are about $9, $57, and $13. It's easy to add $9 to $13, getting $22; then add $22 and $57, getting an approximate total of $79. Or add $57 and $13, for $70, then add $9, for $79.

d) The first item, at two for $15.95, costs about $8 for one. The second item is about $70, and the third about $42. Add $8 and $42, getting $50, and then add $50 and $70, getting the approximate total cost, $120.

a) $12 + $16 + $8 = ($12 + $8) + $16
= $20 + $16 = $36

b) $3.99 + $24.88 + $5.95 = ?
$4 + $25 + $6 = ($4 + $6) + $25
= $10 + $25 = $35 (Approx.)

c) $8.88 + $56.75 + $12.90 = ?
$9 + $57 + $13 = ($9 + $13) + $57
= $22 + $57 = $79 (Approx.)

d) (2 for $15.95) + $69.90 + $41.85 = ?
$8 + $70 + $42 = ($8 + $42) + $70
= $50 + $70 = $120 (Approx.)

Problems – 6

In these problems, you're selling something that you bought earlier at a higher price. See if you can figure in your head how much you've lost. Of course, you've had the use of the item for some time, so it's not a total loss.

a) You're selling a bicycle for $65 that you originally paid $115 for. How much are you losing?

b) You're selling a TV set for $89 that originally cost you $226. What's your loss?

c) You're trying to sell your car for $2,350—a car that set you back $5,925 originally. Your loss?

Loss on Sale

	ITEM	BOUGHT FOR	SOLD FOR	LOSS
a)	Bicycle	One hundred fifteen dollars	Sixty-five dollars	?
b)	TV set	Two hundred twenty-six dollars	Eighty-nine dollars	?
c)	Car	Fifty-nine hundred & twenty-five dollars	Twenty-three hundred & fifty dollars	?

Solutions – 6

These are just subtraction problems, but the numbers are getting a little larger, requiring you to keep track of more in your head.

a) First figure the amount to add to $65 to get to $100 by taking it in steps: first $5, then $30. Then subtract $100 from $115 to see how much further you have to go—that is, $15. Adding $5, $30 and $15, you get $50.

b) Bring the $89 up to $100 by adding $11, then subtract $100 from $226, getting $126. Your loss is the total of $11 and $126, or $137.

c) As usual, work with the number you're subtracting. Bring $2,350 up to a multiple of 100, $2,400, by adding $50. Then add $3,500 to bring the total up to $5,900. You still have $25 to go, so add $50, $3,500, and $25, getting the final answer, $3,575.

a) $$\begin{array}{r} \$115 \\ \underline{-\ 65} \\ ? \end{array} \Rightarrow \begin{array}{rcl} \$65 + \$5 &=& \$70 \\ \$70 + \$30 &=& \$100 \\ \$100 + \underline{\$15} &=& \$115 \\ \$50 \text{ (Loss)} && \end{array}$$

b) $$\begin{array}{r} \$226 \\ \underline{-\ 89} \\ ? \end{array} \Rightarrow \begin{array}{rcl} \$89 + \$11 &=& \$100 \\ \$100 + \underline{\$126} &=& \$226 \\ \$137 \text{ (Loss)} && \end{array}$$

c) $$\begin{array}{r} \$5{,}925 \\ \underline{-2{,}350} \\ ? \end{array} \Rightarrow \begin{array}{rcl} \$2{,}350 + \$50 &=& \$2{,}400 \\ \$2{,}400 + \$3{,}500 &=& \$5{,}900 \\ \$5{,}900 + \underline{\$25} &=& \$5{,}925 \\ \$3{,}575 \text{ (Loss)} && \end{array}$$

Problems – 7

You're thinking of buying some clothes that are on sale. You can solve these problems quickly by getting approximate answers—which is all you'll normally need in this kind of situation. You can, if you wish, solve the problems exactly with only a little more effort.

a) A pair of shoes, regularly $40, is on sale for $31.95. How much would you save approximately? Exactly?

b) A $65 jacket is on sale for $48.88. What would be your approximate savings? Your exact savings?

c) A shirt, normally $22.50, has a sales price of $13.75. What would you save, approximately? Exactly?

Savings on Sale

	ITEM	REGULAR PRICE	SALES PRICE	SAVINGS
a)	Shoes	Forty dollars	Thirty-one ninety-five	Approx. ? Exact ?
b)	Jacket	Sixty-five dollars	Forty-eight eighty-eight	Approx. ? Exact ?
c)	Shirt	Twenty-two fifty	Thirteen seventy-five	Approx. ? Exact ?

Solutions – 7

a) Get an approximate answer by thinking of $31.95 as $32. Subtracting $32 from $40 shows a savings of about $8. The exact answer is obtained by adding a nickel—the difference between $31.95 and $32. So the exact amount saved is $8.05.

b) For an approximate answer, think of $48.88 as $50, and subtract from $65, to get a savings of about $15 (or think of $48.88 as $49 and the approximate savings as $16). Since you added $1.12 to the $48.88 to get $50, add this to the $15, getting an exact amount of $16.12.

c) Think of $22.50 as $23 and $13.75 as $14; the difference is the approximate savings, $9. For an exact answer, since both numbers have been changed, it's better to go back: if the two numbers were $22.50 and $13.50, the difference would be $9. But since the second number is $13.75, the exact amount saved is $0.25 less than $9, or $8.75.

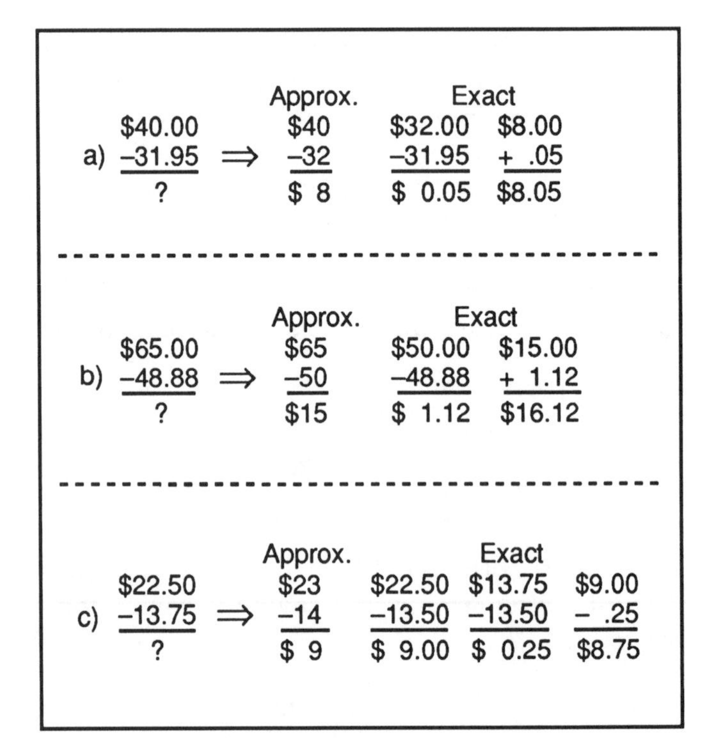

Problems – 8

You've been doing some driving, and are wondering how far you've driven.

a) You've been traveling for 4 hours at an average speed of 40 miles per hour. How far have you driven?

b) You've driven for 3 hours at an average speed of 55 miles per hour. How far have you driven?

c) You've driven for 5 1/2 hours at an average speed of 60 miles per hour. How far have you driven?

d) You've driven for 4 1/2 hours at an average speed of 52 miles per hour. How far have you driven?

Distance Driven

	HOURS DRIVEN	AVERAGE SPEED	DISTANCE
a)	Four hours	Forty miles per hour	?
b)	Three hours	Fifty-five miles per hour	?
c)	Five and one-half hours	Sixty miles per hour	?
d)	Four and one-half hours	Fifty-two miles per hour	?

Solutions – 8

You need to multiply the numbers to find the distance covered.

a) Multiply 4 by 40 to get 160 miles. Or, to multiply by 4, just double 40 twice—twice 40 is 80, and twice 80 is 160 miles.

b) Break up 55 into 50 + 5. Multiplying, you have 50 × 3, which is 150, plus 5 × 3, or 15. The distance is the sum: 150 + 15 is 165 miles.

c) Break up 5 1/2 into 5 + 1/2. Then 5 × 60 is 300, and (1/2) × 60 is 30. The sum is 330 miles. (In this and the previous problem, once you find the first number, you can take one-tenth of it and add.)

d) Break up 4 1/2 into 4 + 1/2. First multiplying 4 × 52, you get 208, and (1/2) × 52 is 26. Now add 208 and 26, getting 234 miles. If you have trouble multiplying 4 × 52, think of 52 as 50 + 2.

a) $$4 \times 40 = 2 \times (2 \times 40) = 2 \times 80 = 160 \text{ miles}$$

b) $$3 \times 55 = (50 + 5) \times 3 = (50 \times 3) + (5 \times 3) = 150 + 15 = 165 \text{ miles}$$

c) $$5\frac{1}{2} \times 60 = \left(5 + \frac{1}{2}\right) \times 60 = (5 \times 60) + \left(\frac{1}{2} \times 60\right) = 300 + 30 = 330 \text{ miles}$$

d) $$4\frac{1}{2} \times 52 = \left(4 + \frac{1}{2}\right) \times 52 = (4 \times 52) + \left(\frac{1}{2} \times 52\right) = 208 + 26 = 234 \text{ miles}$$

Problems – 9

More driving problems. This time you're concerned about gas mileage—that is, how many miles per gallon you've been getting.

a) You've just driven 240 miles and used 12 gallons of gas. How many miles per gallon are you getting?

b) You've driven 315 miles, using 15 gallons of gas. Your miles per gallon?

c) You've driven 432 miles, using 16 gallons of gas. Your miles per gallon?

d) You've driven 319 miles, using 11 gallons of gas. Your miles per gallon?

Gas Mileage

	DISTANCE DRIVEN	GAS USED	MILES PER GALLON
a)	Two hundred forty miles	Twelve gallons	?
b)	Three hundred fifteen miles	Fifteen gallons	?
c)	Four hundred thirty-two miles	Sixteen gallons	?
d)	Three hundred nineteen miles	Eleven gallons	?

Solutions – 9

Division problems are often difficult to do in your head. But these problems can be readily solved if you think of the right way.

a) You need to divide 240 by 12. Since 12 divides into 24 twice, the answer is 20 miles per gallon.

b) You can see the answer to this one if you notice that 15 divides into 300 20 times. Then there is 15 left over, so 15 goes into 315 21 times. You have averaged 21 miles per gallon.

c) Notice that 16 divides into 400 25 times (think of 4/16 as 1/4, which is 0.25), and therefore into 432 27 times. Average: 27 miles per gallon.

d) If you'd gone 330 miles, you'd have averaged exactly 30 miles per gallon; that is, 330 divided by 11 is 30. But you went 11 miles less than 330, so 319 divided by 11 is 29 miles per gallon.

a) $\frac{240}{12} = 20$ miles/gallon

---

b) $\frac{315}{15} = \frac{300}{15} + \frac{15}{15}$

$= 20 + 1 = 21$ miles/gallon

---

c) $\frac{432}{16} = \frac{400}{16} + \frac{32}{16}$

$= 25 + 2 = 27$ miles/gallon

---

d) $\frac{319}{11} = \frac{330}{11} - \frac{11}{11}$

$= 30 - 1 = 29$ miles/gallon

Problems – 10

You live only a mile from work, so you decide to walk. You're wondering how long it will take you, depending on how fast you walk.

a) You think you walk about 4 miles per hour. How long will it take you to walk one mile?

b) If you walk at only 3 miles per hour, how long will it take you?

c) If you walk very fast, at 5 miles per hour, how long will it take you?

d) If you walk at 4 1/2 miles per hour, how long will it take you?

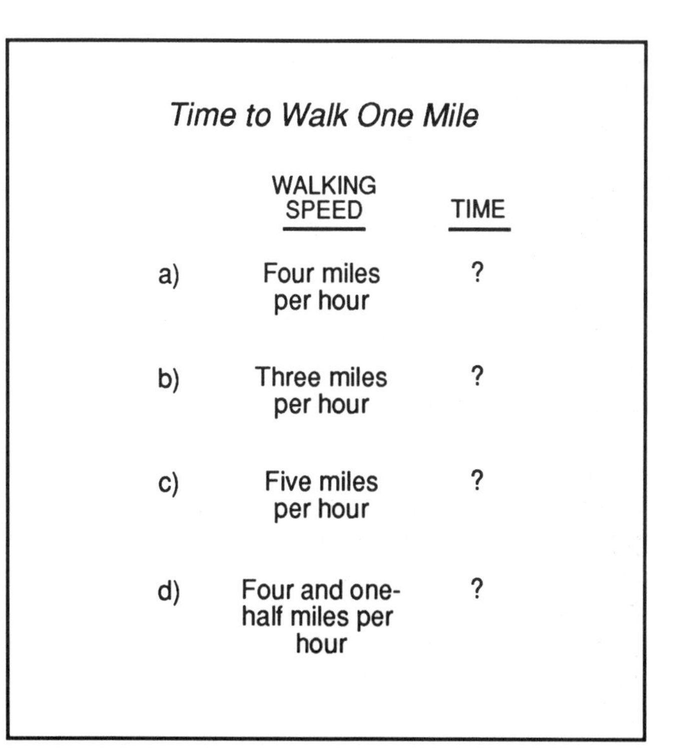

Time to Walk One Mile

	WALKING SPEED	TIME
a)	Four miles per hour	?
b)	Three miles per hour	?
c)	Five miles per hour	?
d)	Four and one-half miles per hour	?

Solutions – 10

a) At 4 miles per hour, it takes you 1/4 of an hour (or 0.25 hour) to walk a mile. In minutes, this is (1/4) × 60, or 15 minutes.

b) At 3 miles per hour, it takes you 1/3 of an hour (or 0.33 hour) to walk a mile. In minutes, this is (1/3) × 60, or 20 minutes.

c) At 5 miles per hour, it takes you 1/5 of an hour (or 0.2 hour) to walk a mile. In minutes, this is only (1/5) × 60, or 12 minutes.

c) At 4 1/2 miles per hour, it takes you 1 ÷ (4 1/2) of an hour to walk a mile. To simplify this, multiply top and bottom of the fraction 1/(4 1/2) by 2, to get 2/9. So it takes 2/9 of an hour (or 0.22 hour) to walk a mile. (That is, the same time it would take you to "walk"—or we should say jog—2 miles at 9 miles per hour). In minutes, this is (2/9) × 60, or 120/9, or (4/3) × 10, or 13.3 minutes.

a) $1 \div 4 = \frac{1}{4}$ hour (or 0.25 hour)

$\frac{1}{4} \times 60 = \frac{60}{4} = 15$ minutes

---

b) $1 \div 3 = \frac{1}{3}$ hour (or 0.33 hour)

$\frac{1}{3} \times 60 = \frac{60}{3} = 20$ minutes

---

c) $1 \div 5 = \frac{1}{5}$ hour (or 0.2 hour)

$\frac{1}{5} \times 60 = \frac{60}{5} = 12$ minutes

---

d) $1 \div 4\frac{1}{2} = \frac{1}{\left(4\frac{1}{2}\right)} = \frac{1 \times 2}{\left(4\frac{1}{2}\right) \times 2} = \frac{2}{9}$ hour (or 0.22 hour)

$\frac{2}{9} \times 60 = \frac{120}{9} = \frac{40}{3} = \frac{4}{3} \times 10 = 13.3$ minutes

Problems – 11

In these problems you want to convert a monthly salary into a yearly salary.

a) Your monthly salary is $800. What is your annual salary?

b) Your monthly salary is $1,500. Your annual salary?

c) Your monthly salary is $1,650. Your annual salary?

d) Your monthly salary is $2,050. Your annual salary?

Monthly/Annual Salary

	MONTHLY SALARY	ANNUAL SALARY
a)	Eight hundred dollars	?
b)	Fifteen hundred dollars	?
c)	Sixteen hundred and fifty dollars	?
d)	Two thousand and fifty dollars	?

Solutions – 11

a) You need to multiply \$800 by 12. Break up 12 into 10 + 2, and multiply separately: 10 × \$800 is \$8,000, and 2 × \$800 is \$1,600. Adding, you find the yearly salary is \$9,600.

b) Break up 12 into 10 + 2, multiply, and add: 10 × \$1,500 is \$15,000, and 2 × \$1,500 is \$3,000. Adding you get \$18,000 as the annual salary.

c) This is tougher, but proceed as before. Break up 12 into 10 + 2, getting \$16,500 plus 2 × \$1,650. Do the latter as 2 × (\$1,600 + \$50), or \$3,200 + \$100, or \$3,300. Add \$16,500 and \$3,300 to get \$19,800 as the annual salary.

d) Let's try a different way. Break up \$2,050 into \$2,000 + \$50, and multiply by 12: (12 × \$2,000) + (12 × \$50), or \$24,000 + \$600, giving an annual salary of \$24,600. Alternatively, break up 12: (10 + 2) × \$2,050 = \$20,500 + \$4,100 = \$24,600.

a) $12 \times \$800 = (10 + 2) \times \$800$
$= (10 \times \$800) + (2 \times \$800)$
$= \$8,000 + \$1,600 = \$9,600$

b) $12 \times \$1,500 = (10 + 2) \times \$1,500$
$= (10 \times \$1,500) + (2 \times \$1,500)$
$= \$15,000 + \$3,000 = \$18,000$

c) $12 \times \$1,650 = (10 + 2) \times \$1,650$
$= (10 \times \$1,650) + (2 \times \$1,650)$
$= \$16,500 + 2 \times (\$1,600 + \$50)$
$= \$16,500 + \$3,200 + \$100$
$= \$19,800$

d) $12 \times \$2,050 = 12 \times (\$2,000 + \$50)$
$= (12 \times \$2,000) + (12 \times \$50)$
$= \$24,000 + \$600 = \$24,600$

Problems – 12

You've been spending different numbers of hours working for two days.

a) Yesterday you worked 3 hours and today you worked 4 1/2 hours. How many hours have you worked in both days?

b) Yesterday you worked 5 1/2 hours and today, 6 1/4 hours. Total hours for the two days?

c) Yesterday you worked 2 3/4 hours and today, 7 1/2 hours. Total hours?

d) Yesterday you worked 12 3/4 hours and today, 9 3/4 hours. Total hours?

Total Hours Worked

	WORKED YESTERDAY	WORKED TODAY	TOTAL
a)	Three hours	Four and one-half hours	?
b)	Five and one-half hours	Six and one-quarter hours	?
c)	Two and three-quarters hours	Seven and one-half hours	?
d)	Twelve and three-quarters hours	Nine and three-quarters hours	?

Solutions – 12

a) Adding 3 and 4, you have 7. Now add 1/2, getting a total of 7 1/2 hours.

b) Add 5 and 6, for 11. Then add 1/2 and 1/4 by realizing that 1/2 is 2/4, and 2/4 + 1/4 is 3/4. The total is therefore 11 3/4 hours.

c) Add 2 and 7, for 9. Add 3/4 and 1/2 by realizing again that 1/2 is 2/4, and 3/4 + 2/4 is 5/4, or 1 1/4. Adding 9 and 1 1/4, you find the total is 10 1/4 hours.

d) Add 12 and 9, for 21. Add 3/4 and 3/4, which is 6/4, or 1 2/4, or 1 1/2. Add 21 and 1 1/2, for a total of 22 1/2 hours.

a) $3 + 4\frac{1}{2} = (3 + 4) + \frac{1}{2} = 7\frac{1}{2}$ hours

---

b) $5\frac{1}{2} + 6\frac{1}{4} = (5 + 6) + \left(\frac{1}{2} + \frac{1}{4}\right)$

$= 11 + \left(\frac{2}{4} + \frac{1}{4}\right) = 11\frac{3}{4}$ hours

---

c) $2\frac{3}{4} + 7\frac{1}{2} = (2 + 7) + \left(\frac{3}{4} + \frac{1}{2}\right)$

$= 9 + \left(\frac{3}{4} + \frac{2}{4}\right) = 9\frac{5}{4} = 10\frac{1}{4}$ hours

---

d) $12\frac{3}{4} + 9\frac{3}{4} = (12 + 9) + \left(\frac{3}{4} + \frac{3}{4}\right)$

$= 21 + \frac{6}{4} = 21 + \left(1 + \frac{2}{4}\right)$

$= 21 + 1\frac{1}{2} = 22\frac{1}{2}$ hours

Problems – 13

With the instant availability of information in computers, sporting events on TV are constantly giving us some numbers about the game. Baseball is a good example.

a) Suppose a batter has been at bat 40 times, and has had 10 hits. What is his batting average—as a fraction, a decimal, a percent?

b) Same questions, but he's had only 7 hits in 56 at-bats.

c) Same questions, but this time he's had 12 hits in 32 at-bats.

d) Same questions, but he's had 40 hits in 180 at-bats.

Baseball Batting Average

	NO. OF AT-BATS	NO. OF HITS	BATTING AVERAGE
a)	Forty	Ten	Fraction ? Decimal ? Percent ?
b)	Fifty-six	Seven	Fraction ? Decimal ? Percent ?
c)	Thirty-two	Twelve	Fraction ? Decimal ? Percent ?
d)	One hundred eighty	Forty	Fraction ? Decimal ? Percent ?

Solutions – 13

a) With 10 hits in 40 at-bats, his fraction of hits is 10/40, which, reduced to lowest terms, is 1/4. As a decimal, this is 0.25—or .250, since in baseball three digits are used, without the initial zero. As a percent, he's hitting 25% of the time.

b) With 7 hits in 56 at-bats, his fraction of hits is 7/56, or 1/8. What is this as a decimal? Since 1/8 is half of 1/4, and 1/4 is 0.25, 1/8 is 0.125. So his batting average is .125. As a percent, it's 12.5%.

c) With 12 hits in 32 at bats, he's hitting 12/32 or (dividing top and bottom by 4) 3/8 of the time. To see what decimal this is, notice that 1/8 is .125 (problem b above); 3/8 is three times this, or .375. The percent is 37.5%.

d) With 40 hits in 180 at-bats, he's hitting 40/180 or (dividing top and bottom by 20) 2/9 of the time. Since 1/9 is .111, this is .222, or 22.2%.

a) $\frac{10}{40} = \frac{1}{4} = .250 = 25\%$

b) $\frac{7}{56} = \frac{1}{8} = .125 = 12.5\%$

c) $\frac{12}{32} = \frac{3}{8} = .375 = 37.5\%$

d) $\frac{40}{180} = \frac{2}{9} = .222 = 22.2\%$

Problems – 14

You're looking at rainfall records for two different years.

a) The annual rainfall for a southern city was 17.0 inches in one year and 14.3 inches in the next. What was the total rainfall for the two years? How much greater was the rainfall in the first year than in the second?

b) The annual rainfall for another city was 22.6 inches in the first year and 18.2 inches in the second. What was the total rainfall? By how much did they differ?

c) The annual rainfall for a third city was 29.3 inches in the first year and 15.8 inches in the second. Total rainfall? Difference?

Annual Rainfall

	RAINFALL IN YEAR 1	RAINFALL IN YEAR 2	TOTAL	DIFFER-ENCE
a)	Seventeen inches	Fourteen point three inches	?	?
b)	Twenty-two point six inches	Eighteen point two inches	?	?
c)	Twenty-nine point three inches	Fifteen point eight inches	?	?

Solutions – 14

a) Think of 17 + 14 as (17 + 3) + 11, which is 20 + 11, or 31. Now add the 0.3, getting 31.3 inches for the total. For the difference, it's not convenient in this problem to try to subtract a multiple of 10. Subtract 15 from 17, getting 2, and add 0.7 (the difference between 15 and 14.3), getting 2.7 inches.

b) Add 22 and 18 as 20 + (2 + 18), or 20 + 20, or 40. Add 0.6 and 0.2, getting 0.8. Then 40 + 0.8 is 40.8 inches. For the difference, subtract 18 from 22, getting 4, and 0.2 from 0.6, getting 0.4. The difference is 4 + 0.4, or 4.4 inches.

c) Add 29 and 15 as (29 + 1) + 14, or 30 + 14, or 44. Add 0.3 and 0.8 to get 1.1. The total is 44 + 1.1, or 45.1 inches. To get the difference, subtract 16 from 29.3 to get 13.3, then add back in 0.2 (the difference between 16 and 15.8): 13.3 + 0.2 = 13.5 inches.

a)

$$\begin{array}{r} 17.0 \\ +\,14.3 \\ \hline ? \end{array} \Rightarrow \begin{array}{l} 17 + 14 = (17 + 3) + 11 = 31 \\ 31 + 0.3 = 31.3 \text{ inches (Total)} \end{array}$$

$$\begin{array}{r} 17.0 \\ -\,14.3 \\ \hline ? \end{array} \Rightarrow \begin{array}{l} 17 - 15 = 2 \\ 2 + 0.7 = 2.7 \text{ inches (Diff.)} \end{array}$$

b)

$$\begin{array}{r} 22.6 \\ +\,18.2 \\ \hline ? \end{array} \Rightarrow \begin{array}{l} 22 + 18 = 20 + (2 + 18) = 40 \\ 0.6 + 0.2 = 0.8 \\ 40 + 0.8 = 40.8 \text{ inches (Total)} \end{array}$$

$$\begin{array}{r} 22.6 \\ -\,18.2 \\ \hline ? \end{array} \Rightarrow \begin{array}{l} 22 - 18 = 4 \\ 0.6 - 0.2 = 0.4 \\ 4 + 0.4 = 4.4 \text{ inches (Diff.)} \end{array}$$

c)

$$\begin{array}{r} 29.3 \\ +\,15.8 \\ \hline ? \end{array} \Rightarrow \begin{array}{l} 29 + 15 = (29 + 1) + 14 = 44 \\ 0.3 + 0.8 = 1.1 \\ 44 + 1.1 = 45.1 \text{ inches (Total)} \end{array}$$

$$\begin{array}{r} 29.3 \\ -\,15.8 \\ \hline ? \end{array} \Rightarrow \begin{array}{l} 29.3 - 16 = 13.3 \\ 13.3 + 0.2 = 13.5 \text{ inches (Diff.)} \end{array}$$

Problems – 15

Here are some problems in percent. You've been shopping, and want to figure the sales tax.

a) You're shopping in a state where the sales tax is 10%, and you buy a $9 item. How much tax would you pay? What would be the total cost to you?

b) The sales tax is 8% and you have chosen a $13 item. How much is the tax? Your total cost?

c) The sales tax is 6% and you are buying an item priced at $3.50. Tax? Total cost?

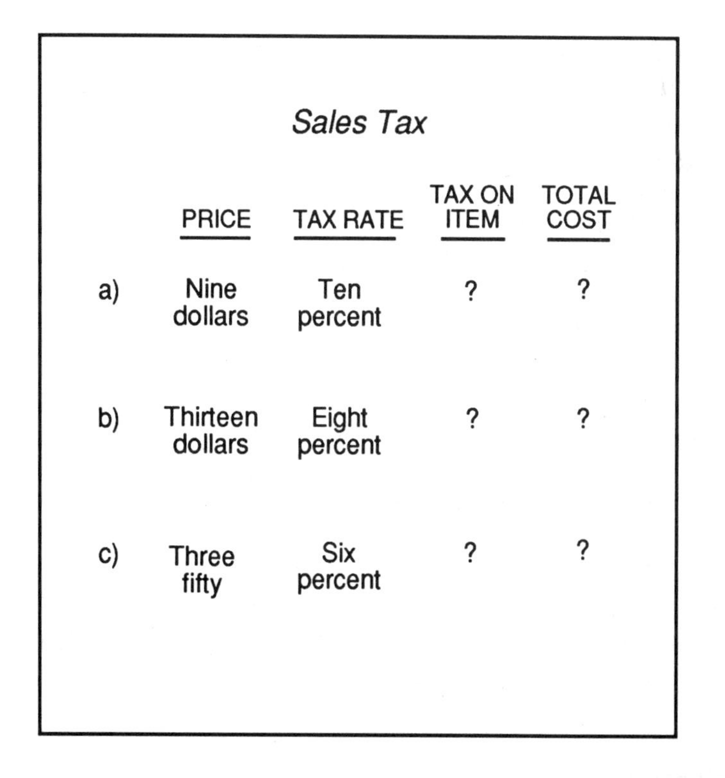

Sales Tax

	PRICE	TAX RATE	TAX ON ITEM	TOTAL COST
a)	Nine dollars	Ten percent	?	?
b)	Thirteen dollars	Eight percent	?	?
c)	Three fifty	Six percent	?	?

Solutions – 15

a) The sales tax is 10%, which is 0.10 of the price. This is $0.10 \times \$9$ or \$0.90 for a \$9 item. The total cost to you is \$9.90.

b) The 8% tax on a \$13 item is $0.08 \times \$13$, which you can multiply by breaking up \$13 into \$10 + \$3, to get $(0.08 \times \$10) + (0.08 \times \$3)$. This is \$0.80 + \$0.24, or \$1.04. The total cost to you is \$13 + \$1.04, or \$14.04.

c) You need to multiply 6%, or 0.06, times \$3.50. Break up the \$3.50 into two parts, \$3 + \$0.50. Now multiply separately and add: $0.06 \times \$3$ is \$0.18—you can see this since 6×3 is 18, and there are two decimal places to account for—and $0.06 \times \$0.50$ is \$0.03. Now add to find the total tax: \$0.18 + \$0.03 is \$0.21. The cost to you is therefore \$3.50 + \$0.21, or \$3.71. What makes this problem more difficult than the others is that it's easy to get confused and lose sight of the decimal point.

a) $0.10 \times \$9 = \$0.90$ (Tax)

$\$9 + \$0.90 = \$9.90$ (Total)

---

b) $0.08 \times \$13 = 0.08 \times (\$10 + \$3)$

$= (0.08 \times \$10) + (0.08 \times \$3)$

$= \$0.80 + \$0.24 = \$1.04$ (Tax)

$\$13 + \$1.04 = \$14.04$ (Total)

---

c) $0.06 \times \$3.50 = 0.06 \times (\$3 + \$0.50)$

$= (0.06 \times \$3) + (0.06 \times \$0.50)$

$= \$0.18 + \$0.03 = \$0.21$ (Tax)

$\$3.50 + \$0.21 = \$3.71$ (Total)

Problems – 16

You're thinking of buying a TV set that is on sale, and would like to figure your savings and the selling price.

a) A set that normally sells for $200 is selling for 10% off. How much would you save? What price would you pay?

b) A $650 set is selling for 20% off. What is your savings? Sales price?

c) A $480 set is selling for 12 1/2% off. What is your savings? Sales price?

TV Set on Sale

	NORMAL PRICE	DISCOUNT	SAVINGS	SALES PRICE
a)	Two hundred dollars	Ten percent	?	?
b)	Six hundred fifty dollars	Twenty percent	?	?
c)	Four hundred eighty dollars	Twelve and one-half percent	?	?

Solutions – 16

a) Your savings would be 10% of \$200, that is, $0.10 \times \$200$, or \$20. Your new price (the sales price) would be \$200 less the savings, or \$180.

b) Your savings would be $0.20 \times \$650$, or \$130—to get this, just think of $0.10 \times \$650$, or \$65, and double it. The new price would be \$650 less your savings, or \$520.

c) What you need is $0.125 \times \$480$, but don't try to do it directly. The problem is easy if you recognize that 12 1/2% or 0.125 is 1/8. Since 1/8 of \$480 is \$60, this is your savings. The new price would be \$480 – \$60, or \$420.

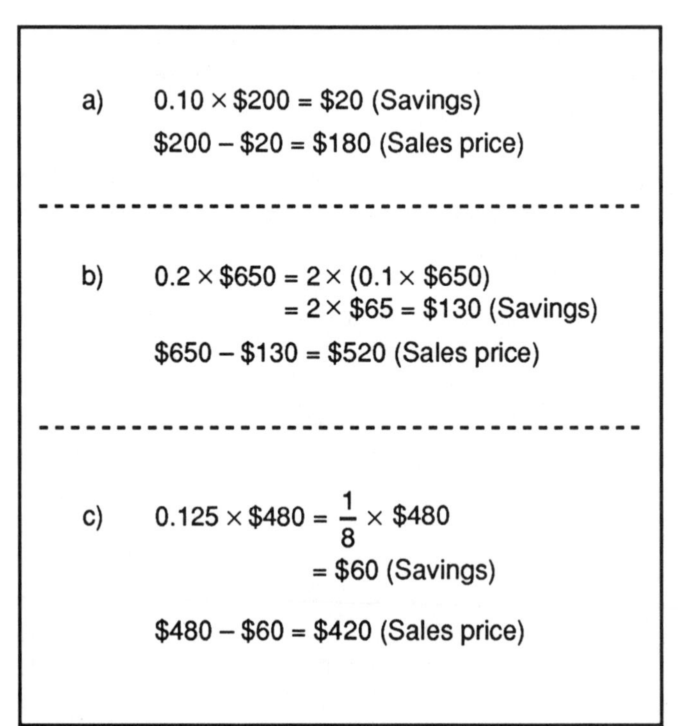

Problems – 17

Here are some problems involving income and rent, and the percent of the former going to the latter.

a) A friend tells you she's spending $300 a month, which is 25% of her income, for rent. What is her monthly income? Her annual income?

b) A different friend tells you he's spending $250 a month, which is 33 1/3% of his income, for rent. What is his monthly income? His annual income?

c) Another acquaintance says she's spending $375 a month, which is 16 2/3% of her income, for rent. What is her monthly income? Her annual income?

Monthly Rent vs. Income

	MONTHLY RENT	PERCENT OF MONTHLY INCOME	MONTHLY INCOME	ANNUAL INCOME
a)	Three hundred dollars	Twenty-five percent	?	?
b)	Two hundred fifty dollars	Thirty-three and one-third percent	?	?
c)	Three hundred seventy-five dollars	Sixteen and two-thirds percent	?	?

Solutions – 17

a) Your friend's rent is 25%, or 1/4, of some number—her monthly income. Multiply her $300 rent by 4 to find her monthly income, $1,200. For her annual income, break up 12: 12 × $1,200 = (10 + 2) × $1,200, which is $12,000 + $2,400, or $14,400. (Or multiply 12 × 12 = 144, and add two zeros).

b) The $250 rent is 33 1/3%, or 1/3, of his monthly income. He must be making 3 × $250, or $750. For his annual income, multiply $750 by 12 to get $9,000 (a simple way: 0.75 is 3/4, and 3/4 of 12 is 9; then add the right number of zeros).

c) Recall that 16 2/3% is 1/6, so her monthly income is 6 × $375. To do this, multiply by 2 to get $750, then multiply by 3 to get $2,250 (break up $750 into $700 + $50 and multiply each by 3 and add: $2,100 + $150). For her annual income, break up 12: 12 × $2,250 = (10 + 2) × $2,250, which is $22,500 + $4,500, or $27,000.

a) 25% = 1/4
4 × $300 = $1,200 (Monthly income)
12 × $1,200 = (10 + 2) × $1,200
= (10 × $1,200) + (2 × $1,200)
= $12,000 + $2,400
= $14,400 (Annual income)

---

b) 33 1/3% = 1/3
3 × $250 = $750 (Monthly income)
12 × $750 = (12 × 3/4) × $1,000
= 9 × $1,000
= $9,000 (Annual income)

---

c) 16 2/3% = 1/6
6 × $375 = 3 × (2 × $375) = 3 × $750
= 3 × ($700 + $50) = $2,100 + $150
= $2,250 (Monthly income)
12 × $2,250 = (10 + 2) × $2,250
= (10 × $2,250) + (2 × $2,250)
= $22,500 + $4,500
= $27,000 (Annual income)

Problems – 18

You're in a store buying some nuts, and wondering if you have enough money to cover the cost.

a) You'd like to buy 1/2 pound of nuts at $3.90 a pound. How much would this cost you?

b) You'd like to buy 1 1/2 pounds at $2.50 a pound. How much?

c) You'd like to buy 3/4 pound at $3.60 a pound. How much?

d) You'd like to buy 2 2/3 pounds at $2.70 a pound. How much?

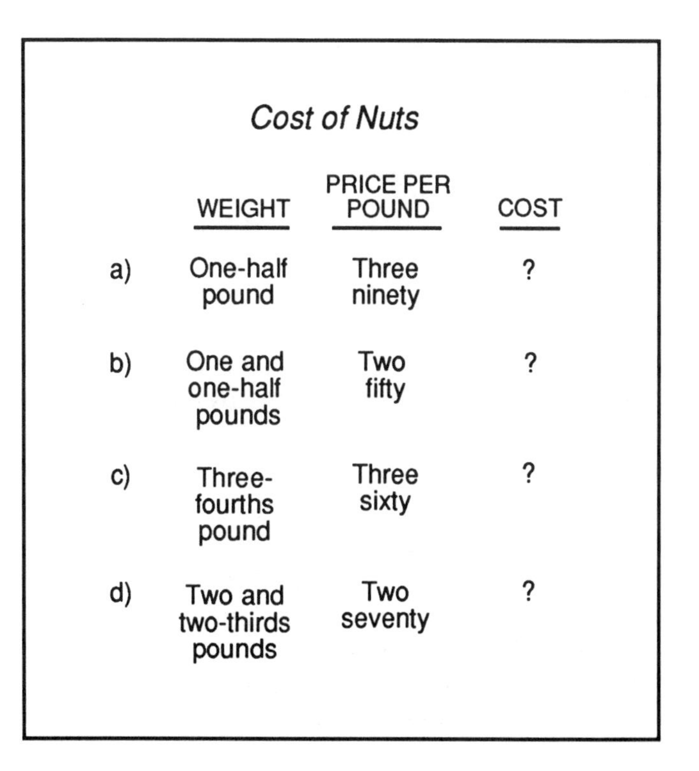

Cost of Nuts

	WEIGHT	PRICE PER POUND	COST
a)	One-half pound	Three ninety	?
b)	One and one-half pounds	Two fifty	?
c)	Three-fourths pound	Three sixty	?
d)	Two and two-thirds pounds	Two seventy	?

Solutions – 18

a) Think of \$3.90 as \$4 – \$0.10. Dividing by 2, you have \$2 – \$0.05, or \$1.95.

b) Since 1 1/2 is 1 + 1/2, or 2/2 + 1/2, or 3/2, you need to multiply by 3 and divide by 2. It's easiest to divide by 2 first: \$2.50 ÷ 2 = \$1.25. Then multiply by 3: 3 × \$1.25 = \$3.75.

c) To find 3/4 of \$3.60, you need to multiply by 3 and divide by 4. Again, it's easiest to divide first: \$3.60 ÷ 4 = \$0.90. Then multiply by 3 to get \$2.70.

d) First change 2 2/3 to 8/3: 2 2/3 = 2 + 2/3, which is 6/3 + 2/3, or 8/3. Now multiply \$2.70 by 8 and divide by 3, as usual dividing first: \$2.70 ÷ 3 = \$0.90, and 8 × \$0.90 = \$7.20. If this last gives you any trouble, think of 8 × \$0.90 as 8 × (\$1 – \$0.10), which is \$8 – \$0.80, or \$7.20.

a) $$\frac{1}{2} \times \$3.90 = \frac{\$3.90}{2} = \frac{\$4 - \$0.10}{2} = \$2 - \$0.05 = \$1.95$$

---

b) $$1\frac{1}{2} \times \$2.50 = \frac{3}{2} \times \$2.50 = 3 \times \frac{\$2.50}{2} = 3 \times \$1.25 = \$3.75$$

---

c) $$\frac{3}{4} \times \$3.60 = 3 \times \frac{\$3.60}{4} = 3 \times \$0.90 = \$2.70$$

---

d) $$2\frac{2}{3} \times \$2.70 = \frac{8}{3} \times \$2.70 = 8 \times \frac{\$2.70}{3} = 8 \times \$0.90 = \$7.20$$

Problems – 19

You're thinking of buying some shoes that are on sale, and you'd like to know what percent you'd save.

a) The shoes are normally $30, and are selling for $20. What percent are you saving?

b) They're normally $40, selling for $34. Percent savings?

b) They're normally $27, selling for $24. Percent savings?

c) They're normally $48, selling for $30. Percent savings?

Buying Shoes

	NORMAL PRICE	SALES PRICE	PERCENT SAVINGS
a)	Thirty dollars	Twenty dollars	?
b)	Forty dollars	Thirty-four dollars	?
c)	Twenty-seven dollars	Twenty-four dollars	?
d)	Forty-eight dollars	Thirty dollars	?

Solutions – 19

a) You're saving $10, the difference between the normal selling price and your price. This is 10/30 or 1/3 of the selling price, or 33 1/3%.

b) You're saving $6 on a $40 pair of shoes, so you're saving 6/40, or 3/20, of the normal price. Divide 2 into 3, getting 1.5, then divide by 10 to get 0.15, or 15%.

c) The $3 you save is 3/27, or 1/9, of the selling price. Since the decimal equivalent of 1/9 is 0.111..., this translates to 11.1%.

d) The $18 that you're saving on a $48 pair of shoes is 18/48 of the selling price. Dividing top and bottom by 6, this reduces to 3/8. The decimal equivalent of 3/8 is 3 × 0.125, or 0.375, so the percent savings is 37.5%.

a) $\$30 - \$20 = \$10$ (Savings)

$$\frac{\$10}{\$30} = \frac{1}{3} = 33\frac{1}{3}\%$$

---

b) $\$40 - \$34 = \$6$ (Savings)

$$\frac{\$6}{\$40} = \frac{3}{20} = 15\%$$

---

c) $\$27 - \$24 = \$3$ (Savings)

$$\frac{\$3}{\$27} = \frac{1}{9} = 11.1\%$$

---

d) $\$48 - \$30 = \$18$ (Savings)

$$\frac{\$18}{\$48} = \frac{3}{8} = 37.5\%$$

Problems – 20

In a restaurant, you've been handed the check and want to leave a tip of about 15%.

a) The total on the check is $8.90. What is your tip?

b) The tab is $26.25. What is your tip?

c) The tab is $13.77. Your tip?

d) The bottom line is $37.48. Your tip?

Leaving a Tip

	AMOUNT OF CHECK	TIP*
a)	Eight ninety	?
b)	Twenty-six twenty-five	?
c)	Thirteen seventy-seven	?
d)	Thirty-seven forty-eight	?

* About fifteen percent

Solutions – 20

In these problems, be sure to round off, as it will save you trouble and not make much difference in the answer. Take 10% (1/10) of the check amount, then half that, for 5%, and add to get 15%.

a) First round $8.90 to $9, then take 10% of that, which is 90¢. Half of 90¢ is 45¢. Add them together to get 90¢ + 45¢ or $1.35 for the tip.

b) Round $26.25 to $26; 10% is $2.60, and half that is $1.30. The tip should therefore be about $2.60 + $1.30, or $3.90 (or about $4).

c) Round $13.77 to $14; 10% is $1.40, and half that is 70¢. Added together for the tip, they come to $1.40 + $0.70, or $2.10.

d) Round $37.48 to $37; 10% is $3.70, and half that is (1/2) × ($4 – $0.30), or $2 – $0.15, or $1.85. Add $3.70 + $1.85 as ($3.70 + $2) – $0.15, or $5.55.

a) $8.90 ⇒ $9
0.10 × $9 = $0.90
$0.90 ÷ 2 = $0.45
$1.35

---

b) $26.25 ⇒ $26
0.10 × $26 = $2.60
$2.60 ÷ 2 = $1.30
$3.90

---

c) $13.77 ⇒ $14
0.10 × $14 = $1.40
$1.40 ÷ 2 = $0.70
$2.10

---

d) $37.48 ⇒ $37
0.10 × $37 = $3.70
$3.70 ÷ 2 = $1.85
$5.55

Problems – 21

You join a new firm and discover that you are one of a number of new hirees. You're told that this represents a certain percent of the size of the firm, and you're wondering how large the firm is.

a) You're one of a group of 33 new hirees. The new group represents a 10% increase in the work force of the firm. What was the original work force? The new work force?

b) You're one of 12 new hirees, which is 15% of the existing work force. What was the original work force? The new work force?

c) You're one of 140 new hirees, which is 12 1/2% of the work force. Original work force? New work force?

Size of Firm

	NUMBER OF NEW HIREES	PERCENT OF WORK FORCE	ORIGINAL WORK FORCE	NEW WORK FORCE
a)	Thirty-three	Ten percent	?	?
b)	Twelve	Fifteen percent	?	?
c)	One hundred forty	Twelve and one-half percent	?	?

Solutions – 21

a) The 33 new employees are 10%, or 1/10, of the work force. Therefore, the work force must have been 10×33, or 330 (alternatively, you can just divide 33 by 0.10 to get this). The new work force is 33 more than the original 330, or 363.

b) The 12 new employees are 15% of some number, which is the original work force. To find that number, divide 12 by 0.15. You can do this by noting that 3 divides each number: 12/15 is 4/5, or 0.8. Now adjust the decimal point to get 80 (this saves the trouble of actually dividing 12 by 0.15). The new work force is 12 more than the original 80, or 92.

c) The 140 new hirees represent 12 1/2%, or 1/8, of the original work force; so multiply 140 by 8—$(100 + 40) \times 8$ is $(8 \times 100) + (8 \times 40)$, or $800 + 320$, or 1,120, the original work force. The new work force is 140 greater, or 1,260.

a)
$$33 \div 0.1 = 33 \times 10$$
$$= 330 \text{ (Original work force)}$$
$$330 + 33 = 363 \text{ (New work force)}$$

---

b)
$$12 \div 0.15 = \frac{12}{15} \times 100 = \frac{4}{5} \times 100$$
$$= 80 \text{ (Original work force)}$$
$$80 + 12 = 92 \text{ (New work force)}$$

---

c)
$$140 \div 0.125 = 140 \times 8 = 8 \times (100 + 40)$$
$$= (8 \times 100) + (8 \times 40)$$
$$= 800 + 320$$
$$= 1{,}120 \text{ (Original work force)}$$
$$1{,}120 + 140 = 1{,}260 \text{ (New work force)}$$

Problems – 22

You have some part-time work coming up, and you're wondering how much you'll be making.

a) You'll be working for 3 hours at $7.50 per hour. How much will you make?

b) You'll be working for 6 hours at $8.25 per hour. What will be your earnings?

c) You'll work for 11 hours at $15.50 per hour. Your earnings?

d) You'll work for 15 hours at $12.20 per hour. Your earnings?

Total Earnings

	TIME WORKED	HOURLY WAGE	EARNINGS
a)	Three hours	Seven fifty	?
b)	Six hours	Eight twenty-five	?
c)	Eleven hours	Fifteen fifty	?
d)	Fifteen hours	Twelve twenty	?

Solutions – 22

a) To multiply 3 by \$7.50, think of \$7.50 as \$7 + \$0.50, and multiply separately: $3 \times (\$7 + \$0.50) = (3 \times \$7) + (3 \times \$0.50)$. The earnings are therefore \$21 + \$1.50, or \$22.50.

b) Think of \$8.25 as \$8 + \$0.25. Multiplying, $6 \times (\$8 + \$0.25) = (6 \times \$8) + (6 \times \$0.25)$. Since $6 \times \$8 = \$48$ and $6 \times \$0.25 = \$1.50$, the earnings are \$48 + \$1.50, or \$49.50.

c) To multiply 11 by \$15.50, break up 11 into 10 + 1. Then $(10 + 1) \times \$15.50 = (10 \times \$15.50) + (1 \times \$15.50)$. The earnings are therefore \$155 + \$15.50, or \$170.50.

d) Multiplying by 15 is usually easy, since you just multiply by 10, then cut that in half (for the 5), then add. So $15 \times \$12.20$ is $(10 \times \$12.20) + (5 \times \$12.20)$, or $\$122 + (1/2 \times \$122)$, or \$122 + \$61, and the earnings are \$183.

a) $3 \times \$7.50 = 3 \times (\$7 + \$0.50)$
$= (3 \times \$7) + (3 \times \$0.50)$
$= \$21 + \$1.50 = \$22.50$

---

b) $6 \times \$8.25 = 6 \times (\$8 + \$0.25)$
$= (6 \times \$8) + (6 \times \$0.25)$
$= \$48 + \$1.50 = \$49.50$

---

c) $11 \times \$15.50 = (10 + 1) \times \$15.50$
$= (10 \times \$15.50) + (1 \times \$15.50)$
$= \$155 + \$15.50 = \$170.50$

---

d) $15 \times \$12.20 = (10 + 5) \times \$12.20$
$= (10 \times \$12.20) + \frac{1}{2} \times (10 \times \$12.20)$
$= \$122 + \left(\frac{1}{2} \times \$122\right)$
$= \$122 + \$61 = \$183$

Problems – 23

You have obtained a loan from your local bank. You borrow a certain amount of money for a specified time, making no payments on the principal—the amount borrowed.

a) What is the interest on a loan of $2,000 for one year at 6% interest?

b) What is the interest on $3,500 for two years at 8% interest?

c) What is the interest on $600 for three years at 5 1/2% interest?

Amount of Interest

	LOAN VALUE	INTEREST RATE	LOAN TIME	INTEREST PAID
a)	Two thousand dollars	Six percent	One year	?
b)	Thirty-five hundred dollars	Eight percent	Two years	?
c)	Six hundred dollars	Five and one-half percent	Three years	?

Solutions – 23

a) To figure the interest, multiply 6%, or 0.06, by \$2,000. Move the decimal point two places to the right for 0.06, and to the left for \$2,000, getting 6 × \$20, or \$120.

b) To multiply 8%, or 0.08, by \$3,500, break up \$3,500 into \$3,000 + \$500. Multiply 0.08 × \$3,000 by moving the decimal point as above, getting 8 × \$30 = \$240. Then multiply 0.08 × \$500 = 8 × \$5 = \$40. Adding, \$240 + \$40 = \$280. You borrowed for two years, so multiply by 2. To do this, think of \$280 as \$300 – \$20, and multiply each term by 2 to get \$600 – \$40, or \$560.

c) To multiply 5 1/2%, or 0.055, by \$600, break up 0.055 into 0.05 + 0.005, and multiply each term separately. Then 0.05 × \$600 is 5 × \$6 or \$30, and 0.005 × \$6 is 1/10th this, or \$3. Adding, \$30 + \$3 is \$33, so the interest for one year will be \$33, and for three years, \$99.

a) $0.06 \times \$2{,}000 = 0.06 \times 100 \times \frac{\$2{,}000}{100}$

$= 6 \times \$20 = \$120$

---

b) $0.08 \times \$3{,}500 = 0.08 \times (\$3{,}000 + \$500)$

$= (8 \times \$30) + (8 \times \$5)$

$= \$240 + \$40 = \$280$ (One year)

$\$280 \times 2 = 2 \times (\$300 - \$20)$

$= (2 \times \$300) - (2 \times \$20)$

$= \$600 - \$40 = \$560$ (Two years)

---

c) $0.055 \times \$600 = \$600 \times (0.05 + 0.005)$

$= (\$6 \times 5) + (\$6 \times 0.5)$

$= \$30 + \$3 = \$33$ (One year)

$\$33 \times 3 = \$99$ (Three years)

Problems – 24

You're wondering about the area of a room, perhaps for the purpose of getting some new furniture or a new carpet.

a) The room is 11 feet by 13 feet. What is the area (in square feet)?

b) The room is 17 feet square. What is the area?

c) The room is 12 1/2 by 14 feet. What is the area?

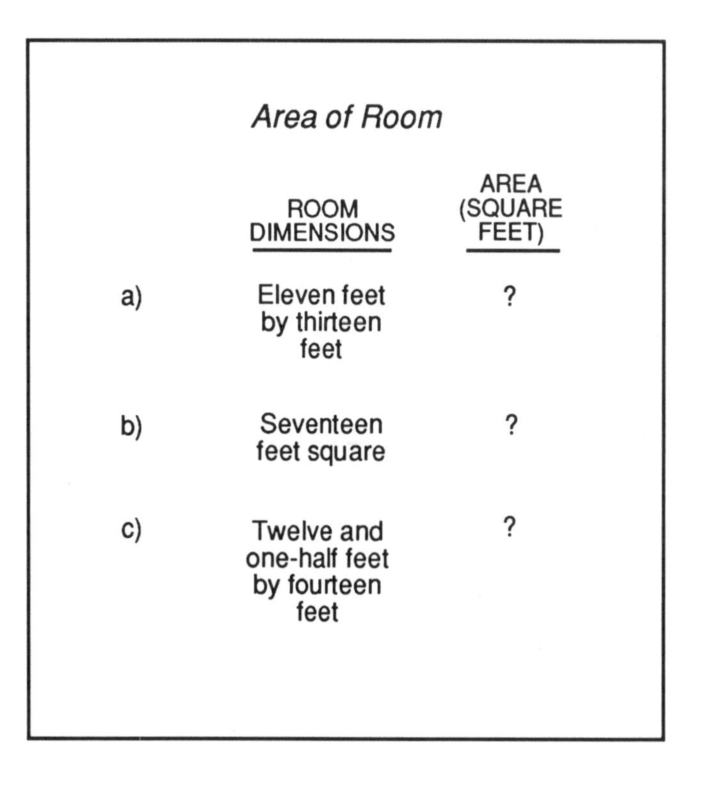

Area of Room

	ROOM DIMENSIONS	AREA (SQUARE FEET)
a)	Eleven feet by thirteen feet	?
b)	Seventeen feet square	?
c)	Twelve and one-half feet by fourteen feet	?

Solutions – 24

a) You can multiply 11 by 13 in either of two ways: (1) Think of 11 as 10 + 1 and multiply separately as $(10 \times 13) + (1 \times 13)$, which is 130 plus 13, or 143 square feet. (2) Remember the identity $(a - 1) \times (a + 1) = a^2 - 1$. With a = 12, you have $11 \times 13 = 12^2 - 1$, which is 144 – 1, or 143 square feet.

b) Use the identity $(a + b)^2 = a^2 + 2 \times a \times b + b^2$, and let a = 10 and b = 7 to get $(10 + 7)^2 = 10^2 + (2 \times 10 \times 7) + 7^2$, which is 100 + 140 + 49, or 289 square feet. If you prefer, let a = 20 and b = –3 and think of 17×17 as $(20 - 3) \times (20 - 3)$, which is $20^2 - (2 \times 20 \times 3) + 3^2$, or 400 – 120 + 9, or 289 square feet.

c) Break up 12 into (10 + 2 + 1/2), and multiply separately, getting $(10 \times 14) + (2 \times 14) + (1/2 \times 14)$, or 140 + 28 + 7, or 175 square feet. Or do this: since 12 1/2 is 100×0.125, or $100 \times (1/8)$, just divide 14 by 8, getting 14/8 or 7/4, or 1.75, and multiply by 100: $100 \times 1.75 = 175$ square feet.

a) $11 \times 13 = (10 + 1) \times 13 = 130 + 13 = 143$ sq ft

or $11 \times 13 = (12 - 1) \times (12 + 1)$
$= 12^2 - 1 = 144 - 1 = 143$ sq ft

---

b) $17 \times 17 = (10 + 7) \times (10 + 7)$
$= 10^2 + (2 \times 10 \times 7) + 7^2$
$= 100 + 140 + 49 = 289$ sq ft

or $17 \times 17 = (20 - 3) \times (20 - 3)$
$= 20^2 - (2 \times 20 \times 3) + 3^2$
$= 400 - 120 + 9 = 289$ sq ft

---

c) $12\frac{1}{2} \times 14 = 14 \times \left(10 + 2 + \frac{1}{2}\right)$
$= (14 \times 10) + (14 \times 2) + \left(14 \times \frac{1}{2}\right)$
$= 140 + 28 + 7 = 175$ sq ft

or $12\frac{1}{2} \times 14 = (0.125 \times 100) \times 14$
$= \left(\frac{1}{8} \times 14\right) \times 100 = \frac{14}{8} \times 100$
$= \frac{7}{4} \times 100 = 1.75 \times 100 = 175$ sq ft

Problems – 25

You're planning to take a driving trip, covering 1,000 miles. You're wondering how long it will take if you drive at different speeds.

a) If you average 50 miles per hour, how long will it take you?

b) If you drive 20% slower, how long will it take you?

c) If you drive 10% faster (than the original 50 miles per hour), how long will it take you?

d) If you drive 10% slower (than the original 50 miles per hour), how long will it take you?

Driving Time for One Thousand Miles

	AVG. SPEED	DRIVING TIME
a)	Fifty miles per hour	?
b)	Twenty percent slower	?
c)	Ten percent faster	?
d)	Ten percent slower	?

Solutions – 25

a) To get the time, divide 1,000 by 50, and you see that it will take you 20 hours (50 into 100 is 2, and add a zero).

b) If your speed is 20% slower, it will be 40 miles per hour; that is, 20% of 50 is 0.20×50 or 1/5 of 50, or 10, and $50 - 10$ is 40. Your new time will be 1,000 divided by 40, or 100/4, or 25 hours.

c) A speed of 50 mph increased by 10% is increased by 5 mph, for a new speed of 55 mph. To divide 1,000 by 55, divide both by 5, getting 200/11. The fractional equivalent of 1/11 is 0.0909..., so 2/11 is 0.1818... Multiply this by 100 to get 18.18, or, rounded off, 18.2 hours.

d) A speed of 50 mph decreased by 10% is 45 mph. To divide 1,000 by 45, divide both by 5, which gives you 200/9. Since 1/9 is 0.111..., 2/9 is 0.222... Multiply this by 100 to get 22.2 hours.

a) $1{,}000 \div 50 = 20$ hours

---

b) $50 - (0.2 \times 50) = 50 - 10 = 40$ mph

$1{,}000 \div 40 = 25$ hours

---

c) $50 + (0.1 \times 50) = 50 + 5 = 55$ mph

$$1{,}000 \div 55 = \frac{200}{11} = 200 \times \frac{1}{11}$$
$$= 200 \times 0.091 = 18.2 \text{ hours}$$

---

d) $50 - (0.1 \times 50) = 50 - 5 = 45$ mph

$$1{,}000 \div 45 = \frac{200}{9} = 200 \times \frac{1}{9}$$
$$= 200 \times 0.111 = 22.2 \text{ hours}$$

3. BACKGROUND: BASIC ARITHMETIC

To count, we use ten digits...

Arithmetic begins with counting. All of our numbers are made up of only ten numerals, or digits: 0,1,2,3,4,5,6,7,8,9 (so-called "Arabic numerals," as they were inherited by us from India via the Arabs). We can get by with only ten digits because we have a *place-value* system: a digit has a different value, depending on its place in the number. For example, the 3 in 13 has a different value than it does in 36; in 13 it means three ones or units, and in 36 it means three tens. And because it means three *tens* and not three of some other number, our system is called a *decimal* system: it uses ten digits and counts with a *base* of ten.

The names of numbers after ten—except for eleven and twelve, which take time to explain—clearly reflect their base of ten: thirteen (literally "three and ten"), fourteen ("four and ten"), on up to twenty ("two tens"), thirty ("three tens"), and so on. Why base ten? Because we have ten fingers.

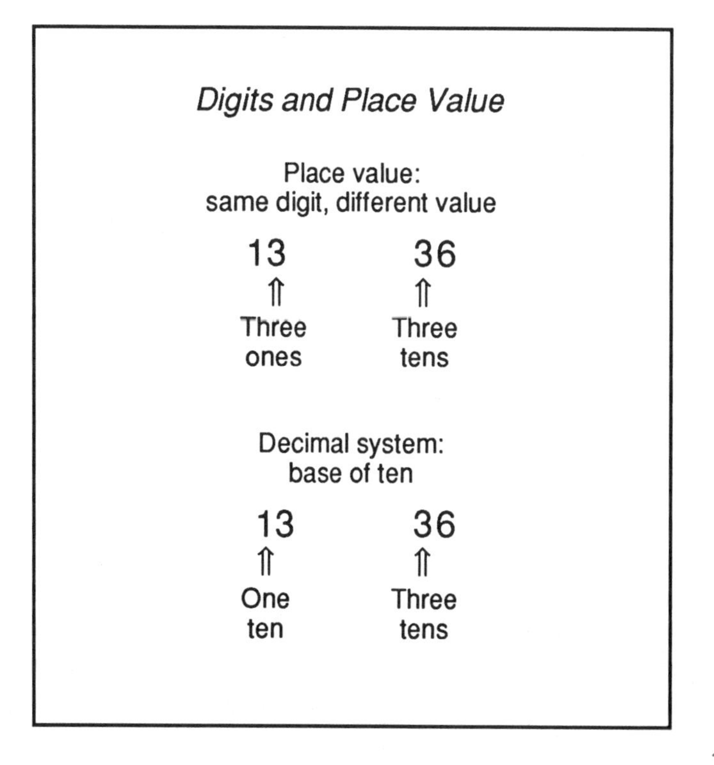

to write numbers of any size.

Our number system uses bunches of different size—all based on the number ten—to represent numbers. First are ones or units (1), then a bunch size of ten (10 × 1), then a larger bunch of a hundred (10 × 10), then a still larger bunch of a thousand (10 × 10 × 10), and so on. Each bunch is ten times larger than the previous one. Let's see how this works in writing numbers. A number like 274 means two hundreds plus seven tens plus four ones: (2 × 100) + (7 × 10) + (4 × 1). If a bunch size is missing, it's handled with a 0, which serves as a "place holder." Without that key digit 0, there would be no good way to distinguish between, say, 274 and 2,074—the latter being two thousands plus *zero* hundreds plus seven tens plus four ones: (2 × 1,000) + (0 × 100) + (7 × 10) + (4 × 1).

This scheme works for any number. Using place value, larger and larger bunches, and the ten digits 0–9 we can write as large a number as we like.

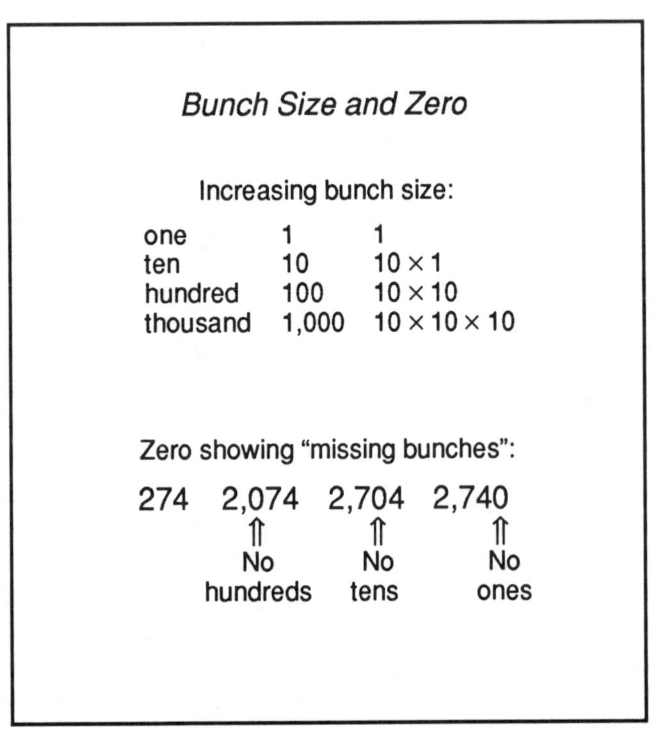

Addition and subtraction are just fast counting.

When you add two numbers—we're talking about *whole numbers*, or *integers*, not fractions, which are described later—you're basically counting. If you add 4 and 3, you're asking what number you'd reach if you combined 4 items with 3 items, and counted all of them. If you know the answer—the *sum*, in this case 7—it saves you the trouble of counting. But if you forget, you can count.

Similarly, if you subtract 4 from 7, you're asking what's left if you remove a group of 4 items from the group of 7. To find out, you can count the remaining items and see. Again, knowing the answer—the *difference*—saves you from having to count. The addition 4 + 3 = 7 yields two subtractions: 7 – 4 = 3 and 7 – 3 = 4.

Subtraction is the reverse of addition; you can answer a subtraction problem such as 7 – 4 by asking what number must be added to 4 to give 7.

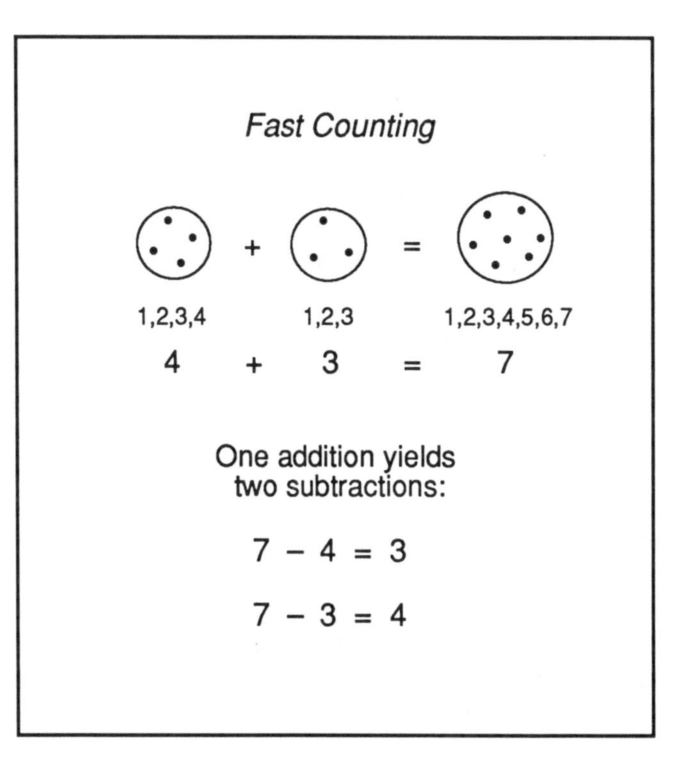

Multiplication is repeated addition,...

When you multiply two whole numbers, say 3 and 2, you're combining 3 groups of 2 each into a single group. The result of multiplying—in this case 6—is called the *product*, and 6 is a *multiple* of 2 and 3. Multiplication is repeated addition: 3×2 is the same as $2 + 2 + 2$. Because of this, it saves a lot of time. This is not so obvious with small numbers like 3×2, but suppose you had to add a column of 46 numbers, each of which is 57. This would take you a while, and chances are you'd make an error somewhere. But you can multiply rapidly, using pencil and paper or a calculator with this one, as it's not the sort you can easily do in your head.

Since addition is fast counting, multiplication, being repeated addition, is *very* fast counting. If you had to *count* the items in 46 groups of 57 each (or 57 groups of 46 each, which has the same count, as discussed shortly), you'd soon give up and find better things to do.

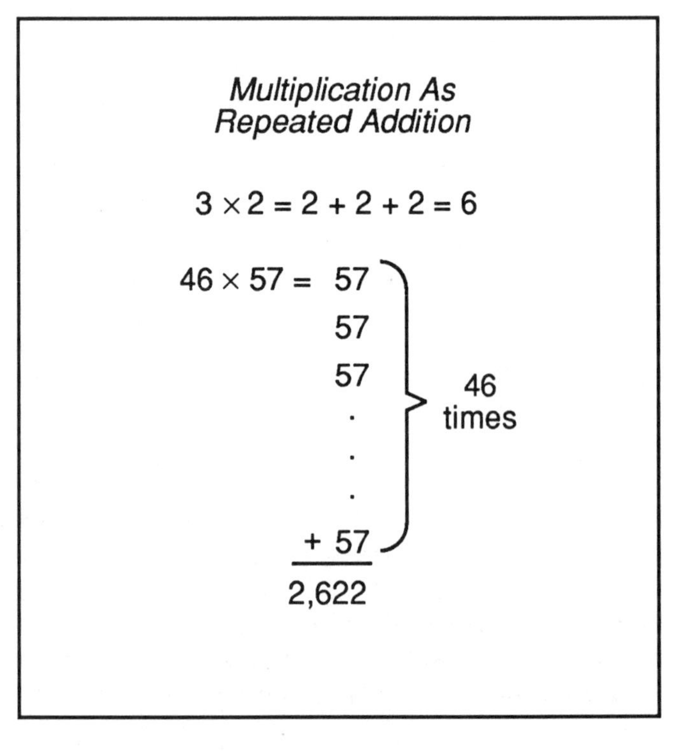

and division is repeated subtraction.

In division, a large group is broken into smaller groups of equal size. To divide 6 by 2, break up the group of 6 into 2 groups of 3 each; 6 divided by 2 is 3, where 3 is called the *quotient*. To divide, you can do repeated subtraction. If you keep subtracting 2 from 6 until there's nothing left, you find there are 3 subtractions. Let's take a larger number: from the previous page we know that $46 \times 57 = 2{,}622$. So you could subtract 57's until you reach zero. The number of 57's you take away is the quotient, 46. You don't need to subtract the 57's one at a time; you can do it in bunches. But you do need to keep track of how many.

Just as one addition leads to two subtractions, a single multiplication yields two divisions: $3 \times 2 = 6$ means $6 \div 3 = 2$ and $6 \div 2 = 3$. And as subtraction is the reverse of addition, division is the reverse of multiplication. To divide 6 by 2, you can ask what number you must multiply 2 by to get 6.

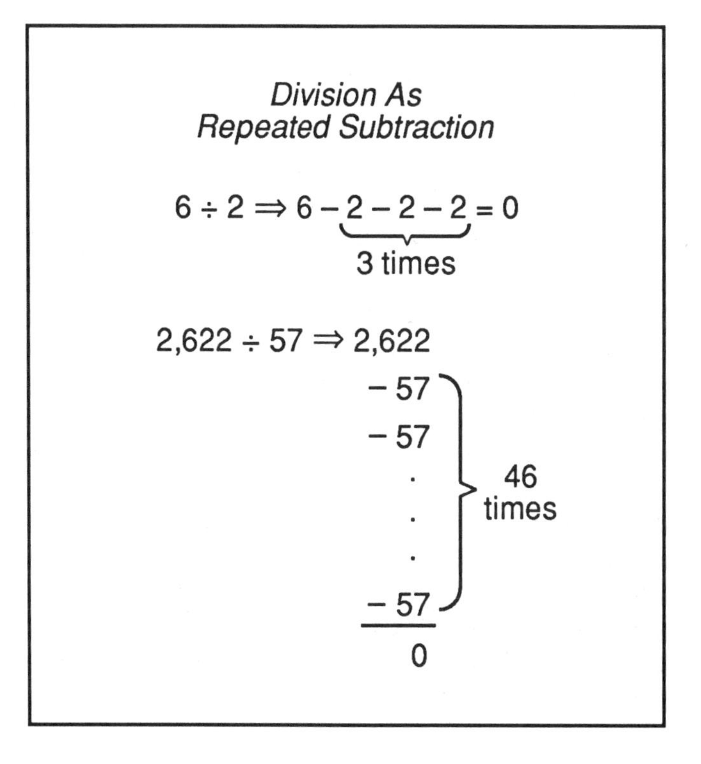

If you forget the "tables," try counting.

When younger you learned—perhaps memorized—the addition and multiplication tables. But the tables aren't just a lot of isolated facts, and those who memorized them may have overlooked many of the relations between the numbers. If you compare 9 × 7 with 3 × 7, you'll see that the former is three times the latter; 4 + 5 is the same as 3 + 6; 6 × 2 = 4 × 3; and so on. If you forget something, think of other ways to get there—other relations between numbers—or just try counting.

Let's say you can't remember what 8 + 6 is. Just think of a different grouping of the 8 + 6 items as 10 + 4. You'll know what this is, 14 (that's the *meaning* of 14—10 plus 4). Another way: lay out two lengths, 8 and 6, and redraw them as 10 + 4—again, 14. Getting in the habit of regrouping and rearranging can be useful in adding numbers in your head; for example, you can think of 80 + 60 as 100 + 40.

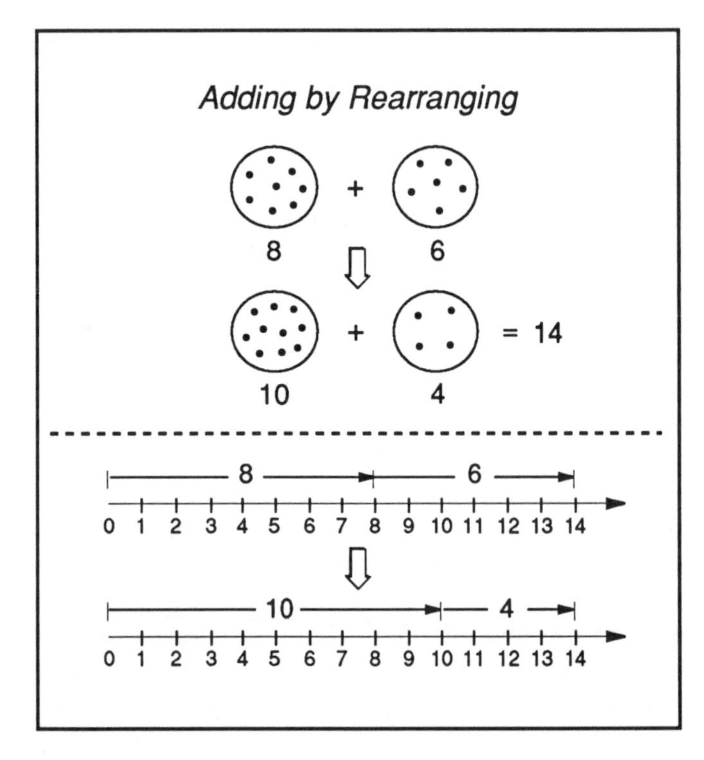

You're free to add or multiply in any order...

Numbers can be added in any order. In adding, you're combining groups into one larger group and counting the result. Clearly, how you combine them doesn't matter; the total count will be the same. So, for example, $3 + 6 = 6 + 3$. Numbers can also be multiplied in any order: $3 \times 6 = 6 \times 3$. This will be clear if you form two arrays, one with 3 rows and 6 columns and the other with 6 rows and 3 columns. You can see that they have the same number of dots—in fact, the second array is just the first rotated to a vertical position.

These relations are examples of the commutative laws of addition and multiplication. Because of them, the number of entries in the addition and multiplication tables is essentially cut in half.

In subtraction and division, order does matter: $6 - 3$ is not equal to $3 - 6$, nor is $6 \div 3$ equal to $3 \div 6$.

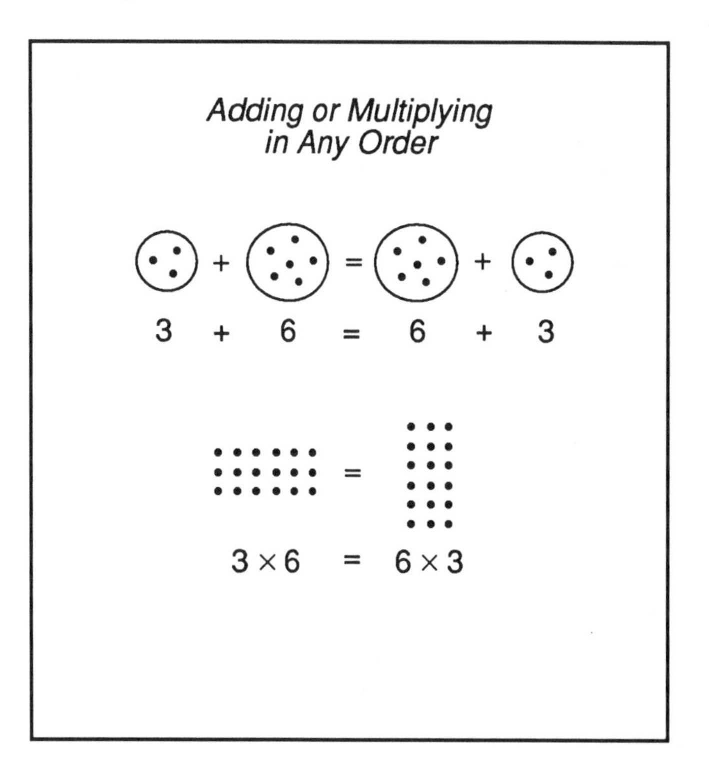

and in any groups.

In a string of numbers to be added, parentheses show which numbers are to be added first: $(46 + 37) + 4$ means add 46 and 37, then add 4. You can move parentheses around and group in any way: for example, $(46 + 37) + 4 = 46 + (37 + 4)$. Since you can also reorder numbers (previous page), you can rewrite this as $(46 + 4) + 37$, which is $50 + 37$; this is easier to do in your head than $(46 + 37) + 4$. You have the same freedom in multiplying; group the numbers any way you like. For example, $(15 \times 23) \times 2 = 15 \times (23 \times 2)$. Since you can reorder the numbers here also, rewrite this as $(15 \times 2) \times 23$—easier to do mentally than $(15 \times 23) \times 2$. Multiplying out, $(15 \times 2) \times 23 = 30 \times 23$, or 690.

The regrouping relations are examples of the associative laws of addition and multiplication. These together with the commutative laws (previous page) let you rearrange problems in various ways to make them easier to handle.

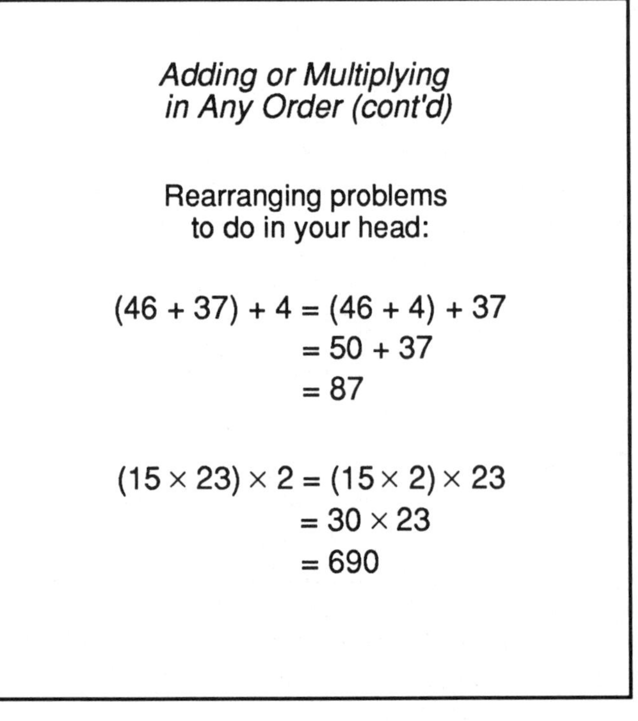

It's possible to add without "carrying"...

In doing addition, people sometimes have trouble with "carrying." When you add 54 and 87, 4 + 7 causes an overflow; 11 has two digits and there's room for only one digit in the ones column. You probably were taught to handle this overflow by writing a small 1 (or other digit) above left of the ones column as a reminder that the 11 you got in adding is 1 ten plus 1 one.

You can avoid carrying if you wish, whether you're doing the problem on paper or in your head. On paper, add each column separately: 4 + 7 is 11, and 5 + 8 is 13—or, really, 50 + 80 is 130. Then just add. What you're doing is 54 + 87 = (50 + 4) + (80 + 7) = (50 + 80) + (4 + 7) = 130 + 11. If you're doing the problem in your head, change 87 to a multiple of 10—in this case 90—by adding 3, and subtract 3 from 54 to make 51. Now change 90 to 100 by adding 10, and 51 to 41 by subtracting 10, and the problem all but solves itself.

"Carrying" in Addition

Adding with carrying:

$$\begin{array}{r} {}^{1}\\[-6pt] 54 \\ +\,87 \\ \hline 141 \end{array}$$

Adding without carrying:

On paper	In your head
$\begin{array}{r} 54 \\ +\,87 \\ \hline 11 \\ 13 \\ \hline 141 \end{array}$	$\begin{array}{r} 54 \\ +\,87 \\ \hline \end{array} \Rightarrow \begin{array}{r} 51 \\ +\,90 \\ \hline \end{array} \Downarrow \begin{array}{r} 41 \\ +\,100 \\ \hline 141 \end{array}$

and subtract without "borrowing."

Subtraction often requires the reverse operation of carrying, where you must "borrow" from the left column. All you have to do is to keep track of the bunch size to which each digit applies, and subtract bunches of similar size.

Say you're trying to subtract 48 from 74. Think of 74 not as 7 tens and 4 ones, but as 6 tens and 14 ones; that's what you're doing when you borrow. The problem is now simple, since you're subtracting 4 tens and 8 ones (that is, 48) from 6 tens and 14 ones. The answer: 2 tens and 6 ones, or 26.

In subtracting in your head, you can often avoid borrowing by rearranging numbers. Try to make the number being subtracted a multiple of 10. In the above problem, you can change 48 to 50 and, to compensate, change 74 to 76, and get the same answer when you subtract. Now there is no borrowing to worry about: 76 – 50 = 26.

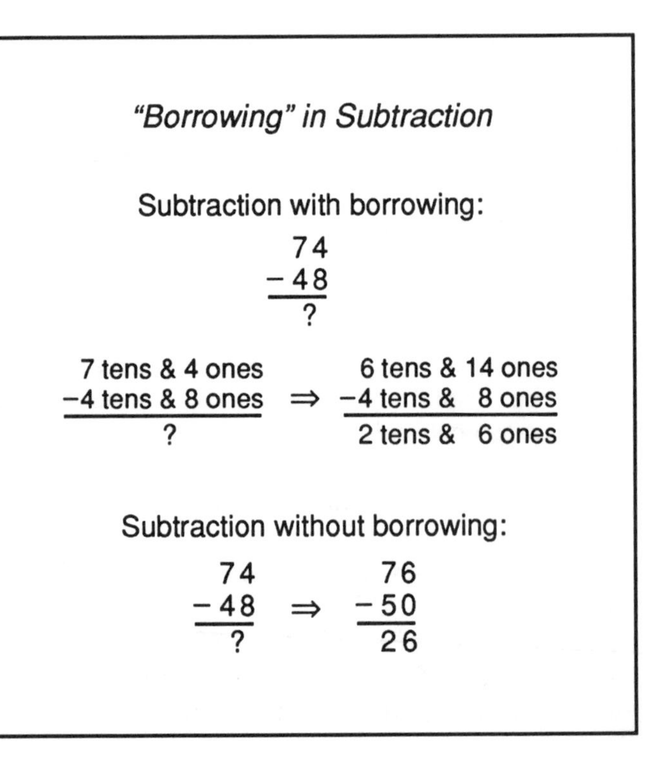

Multiplication is done in stages,...

The simple multiplication tables you learned in school go up to 9 × 9—as high as you can go with single-digit numbers. What happens when you go higher?

Suppose one of the numbers is a two-digit number: say, 8 × 24. If you multiply this out in the normal way, you'll first multiply 8 × 4, which is 32. You'll write down the 2 and carry the 3. Then you'll multiply 8 × 2, which is 16, and add the carried 3 to get 19. Final answer: 192. This standard procedure tends to mask what you're really doing, which you can see if you look at the products as you multiply: 8 × 4 is 32, and 8 × 2 is really 8 × 20, which is 160. Then 160 + 32 is 192. To see this more clearly, it helps to draw a rectangle, where one side is 8 and the other side is 24. Think of 24 as 20 + 4 (which it *is*, of course), and then multiply each part separately by 8, getting (8 × 20) + (8 × 4), which is 160 + 32, or 192.

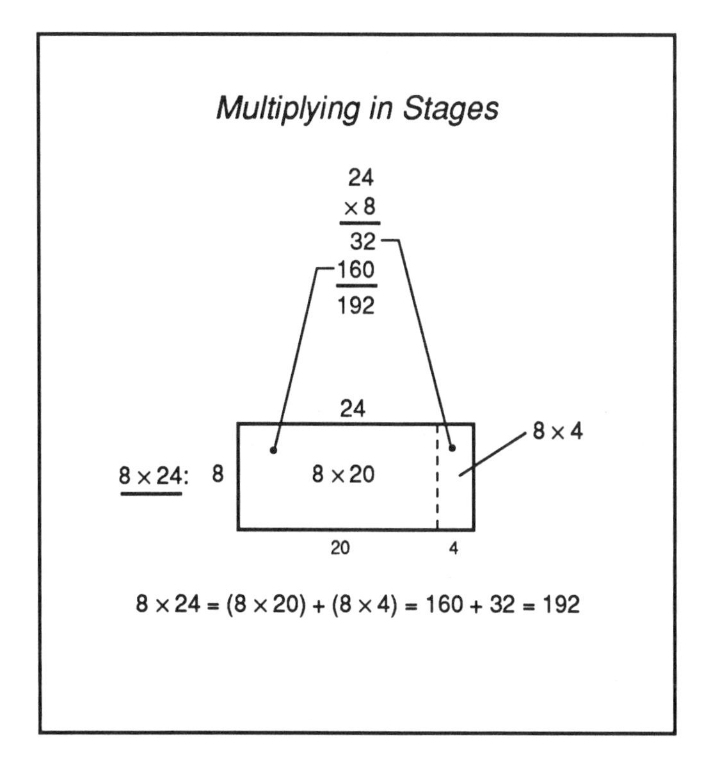

making use of "partial products."

If both numbers have two digits, things get more complicated. Let's look at 13×27. Again, the usual paper-and-pencil procedure hides what's going on; when you say "$1 \times 2 = 2$," you may forget you're multiplying 10×20, and getting 200. If you write this problem as $(10 + 3) \times (20 + 7)$ and draw a diagram, you'll see there are four parts (in the diagram on the previous page there were only two). These parts, or "partial products," are: 10×20, 10×7, 3×20, and 3×7. Adding them separately, $200 + 70 + 60 + 21 = 351$. If you're doing this in your head, it's easy to forget one of the four necessary parts. So when multiplying two two-digit numbers, it's better to do it in two parts. In this example, you could do it as $(10 + 3) \times 27$, which is $(10 \times 27) + (3 \times 27)$, or $270 + 81$, or 351.

What do you do with more than two digits? Except for special cases, use pencil and paper, a calculator, or approximate numbers.

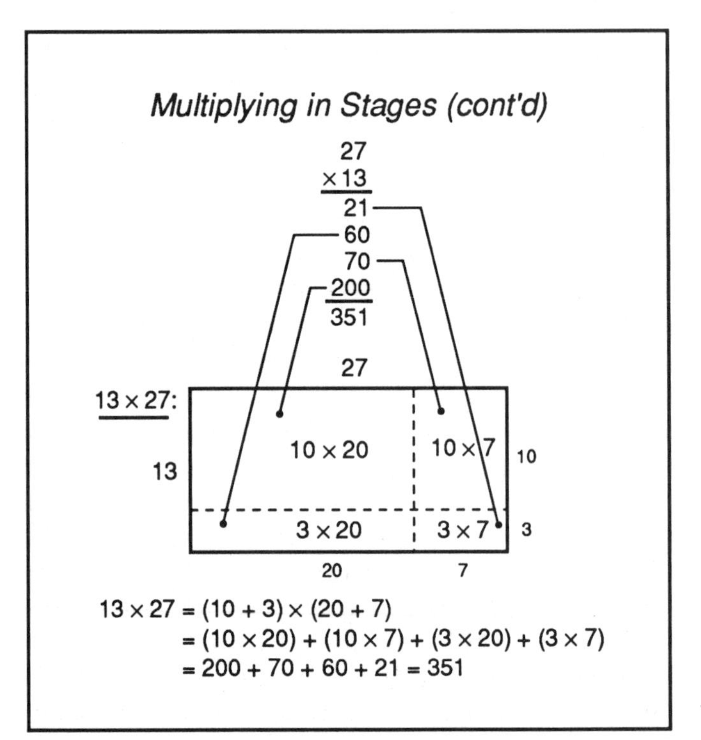

In division, good guesses help,...

While division is repeated subtraction, that's not the way division problems are usually solved. But thinking in terms of subtraction will help you make better guesses when doing division—and it beats "long division," which isn't suitable for doing in your head.

The better your guesses, the faster you reach the answer. Suppose you want to divide 312 by 13. You start by guessing there are ten 13's in 312, but on subtracting these (130), you find you still have 182 left. You take another ten 13's, and have 52 left. Taking four more, you have your answer: 10 + 10 + 4, or 24. If you had taken twenty 13's to begin with, you'd have gotten the answer faster. Could you have guessed this? Well, you know that 20 × 13 is 260, which is less than 312, and 30 × 13 is 390, which is too much. So 20 would be a good first guess. Mentally subtracting 260 from 312, you see you need four more 13's, for a total of 24.

Guessing in Division

$$312 \div 13 = ?$$

$$\begin{array}{rl} 312 & \\ -130 & (10 \times 13) \\ \hline 182 & \\ -130 & (10 \times 13) \\ \hline 52 & \\ -\ 52 & (\ 4 \times 13) \\ \hline 0 & \underline{\quad} \\ & 24 \end{array}$$

$$312 \div 13 = 24$$

but there may be a remainder.

Often a number will not divide evenly into another: 2 does not go nicely into 7, for example. There is a "remainder": 7 ÷ 2 gives a quotient of 3 and a *remainder* of 1.

Let's say you want to divide 245 by 17. As mentioned, in division, you'll arrive at the correct answer even with bad guesses, but you don't want to overshoot. If your initial guess is too large, you'll have to backtrack. Suppose that in this problem you start by trying to subtract twenty 17's. You'll soon discover you've gone too far; 20 × 17 is 340, which is greater than 245. So back up and guess to subtract ten 17's (or 170). Mentally subtracting 170 from 245 leaves 75. You can't take away five 17's (which is 85—more than 75 and therefore too much), so try four 17's. This is 68, which subtracted from 75 leaves 7. Since you can't subtract any more 17's, the answer (quotient) is 10 + 4, or 14, with a remainder of 7.

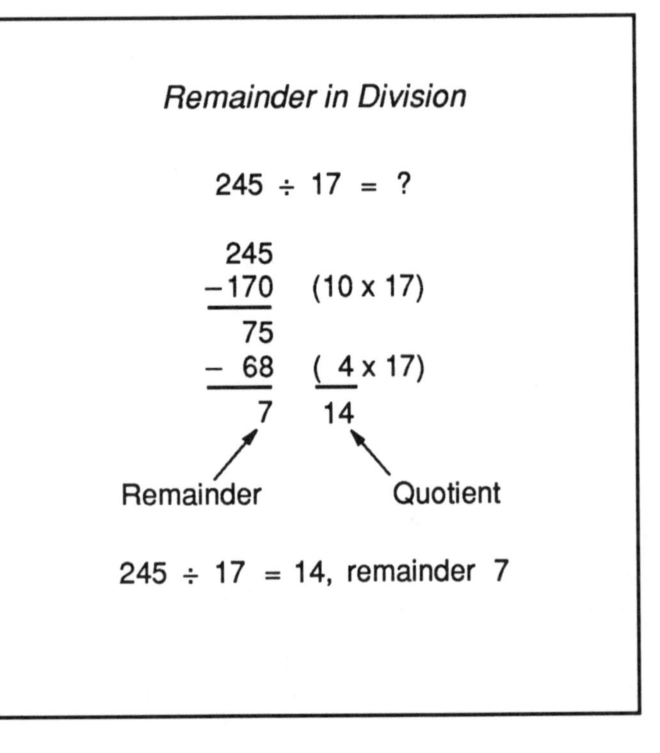

Numbers can be split for multiplying...

Multiplication and addition are related in a useful way. Look at 3×2 and 4×2 and compare the sum of these with 7×2. The sum of the first two, $6 + 8$, is 14—the same as 7×2. Why is this? Think of 3×2 as three 2's and 4×2 as four 2's; together they make seven 2's. This relation (the distributive law) works for any numbers: 8×6 can be changed to $(5 + 3) \times 6$, which is $(5 \times 6) + (3 \times 6)$.

Let's look at some larger numbers: A problem like 12×14 can be broken up in different ways—$(5 + 7) \times 14$, $12 \times (8 + 6)$, and so on—but the first that comes to mind (you'd like to multiply by 10) is to think of it as $(10 + 2) \times 14$. Then multiply these separately, as $(10 \times 14) + (2 \times 14)$. This is easy to do in your head: $140 + 28$, or 168.

A different problem may call for subtracting rather than adding: $18 \times 23 = (20 - 2) \times 23$, which becomes $(20 \times 23) - (2 \times 23)$, or $460 - 46$, or 414.

Multiplying by Breaking Up

$$
\begin{aligned}
8 \times 6 &= (5 + 3) \times 6 \\
&= (5 \times 6) + (3 \times 6) \\
&= 30 + 18 = 48
\end{aligned}
$$

$$
\begin{aligned}
12 \times 14 &= (10 + 2) \times 14 \\
&= (10 \times 14) + (2 \times 14) \\
&= 140 + 28 = 168
\end{aligned}
$$

$$
\begin{aligned}
18 \times 23 &= (20 - 2) \times 23 \\
&= (20 \times 23) - (2 \times 23) \\
&= 460 - 46 = 414
\end{aligned}
$$

or dividing,...

The same idea of breaking up numbers applies if you're dividing. First let's look at some numbers that are easy to work with: $6 \div 3$ and $9 \div 3$. The sum of the two divisions, $2 + 3$, or 5, is the same as if you'd divided $6 + 9$, or 15, by 3. So what was true in multiplying also holds in dividing: breaking up numbers can be helpful in changing problems into simpler ones. Suppose you're thinking about $345 \div 15$. Break up 345 into $300 + 45$, and then divide: $300 \div 15 = 20$ and $45 \div 15 = 3$, so $345 \div 15 = 20 + 3$, or 23. If you aren't sure about $300 \div 15$, break the numbers down further: $345 = 150 + 150 + 45$. Now divide each of these three terms separately by 15: $10 + 10 + 3 = 23$.

As in multiplying, instead of rearranging by addition you can use subtraction if the problem calls for it. With $252 \div 14$ you might think of 252 as $280 - 28$ and divide each term, getting $(280 \div 14) - (28 \div 14)$, or $20 - 2$, or 18.

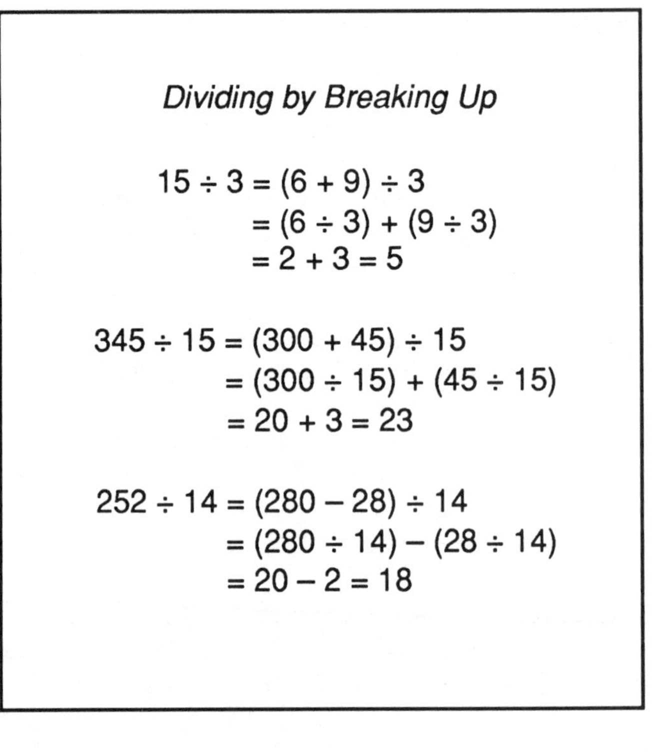

or they can be factored into smaller parts.

Some numbers, such as 2, 3, 5, etc., are not divisible by any other numbers. These are *prime* numbers. Other numbers, such as 4, which is 2×2, or 12, which is 4×3, are *composite*; they can be broken down into their *factors*. The factors may be prime or not, as the example of 12 shows. If you notice that a number is composite, you may find it easier to break it into its factors before multiplying.

For example, you want to multiply 16×35. You can leave 16 alone, and break up 35 into 5×7. Then multiply: 16×5 is 80, and 80×7 is 560. Alternatively, you may make use of the fact that $16 = 2 \times 8$. So you double 35, getting 70, then multiply by 8 to get 560. You could have broken down further since $8 = 2 \times 2 \times 2$, but this wouldn't help.

Factoring is also useful in division: to divide 336 by 21, note that $21 = 3 \times 7$. Divide 336 by 3 to get 112, then divide 112 by 7 to get the answer, 16.

Breaking into Factors

$$\begin{aligned} 16 \times 35 &= 16 \times (5 \times 7) \\ &= (16 \times 5) \times 7 \\ &= 80 \times 7 = 560 \end{aligned}$$

or

$$\begin{aligned} 16 \times 35 &= (2 \times 8) \times 35 \\ &= (2 \times 35) \times 8 \\ &= 70 \times 8 = 560 \end{aligned}$$

$$\begin{aligned} 336 \div 21 &= 336 \div (3 \times 7) \\ &= (336 \div 3) \div 7 \\ &= 112 \div 7 = 16 \end{aligned}$$

The easiest number to work with is 10.

Since our decimal system is based on 10, it and its multiples and powers are the easiest numbers to work with. A multiple of 10 is 20, 30, etc.—any number multiplied by 10. A power of 10 is 10 multiplied by itself: 10×10 is 100 or 10^2; $10 \times 10 \times 10$ is 1,000 or 10^3, and so on. Here 2 and 3 are *powers,* or *exponents*, which save space.

We've seen (p. 82) that the number 274 has the meaning $(2 \times 10^2) + (7 \times 10) + (4 \times 1)$, where we've used 10^2 for 100. If 274 is multiplied by 10, it becomes $(2 \times 10^3) + (7 \times 10^2) + (4 \times 10) + (0 \times 1)$, or 2,740. The result is just 274 with a zero attached to the right end. Multiplying by 10 again—that is, multiplying by 100, or 10^2—produces 27,400; two zeros are added to the end of 274. To multiply by 1,000 (10^3), add three zeros. And so on: to multiply by a power of 10, add as many zeros on the right as are in the power of 10. (We'll postpone dividing by powers of 10 until later.)

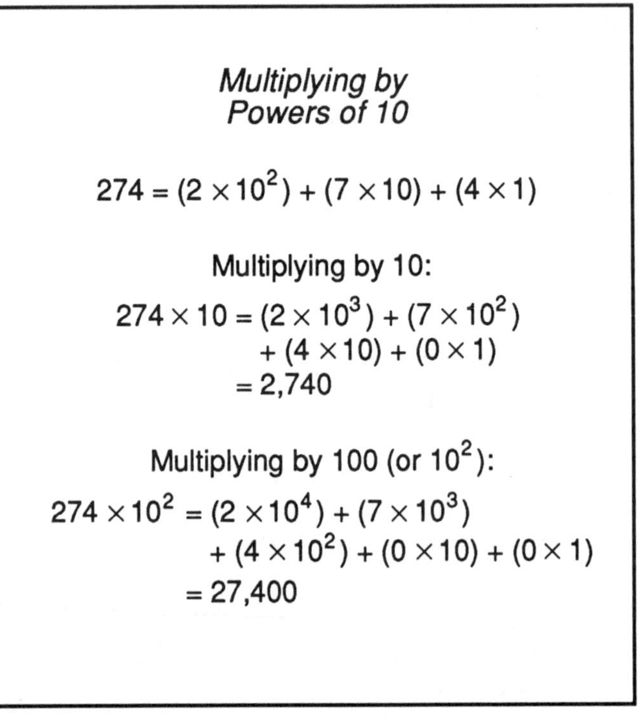

Fractions are parts of a whole,...

In telling time we say "a quarter of three" or "half past seven." We buy a half-dozen eggs or half-gallon of milk, and pay for them with quarters and half-dollars. Fractions are a part of our language. Anything can be divided into parts; whatever is being divided is considered a whole. A pie may be cut in half, making two halves. One of the halves is designated by the fraction 1/2, with 1 called the *numerator* and 2 the *denominator*. The 2 indicates how many parts there are, and the 1 denotes one of those parts. Dividing the pie once more produces four parts. Each part is one-fourth, or 1/4, of the pie. Three parts are three-fourths, or 3/4. The whole pie is four fourths (4/4) or two halves (2/2): 2/2 = 4/4 = 1.

Instead of a pie, let's cut a rectangular cake into three equal pieces. One piece is one-third (1/3) of the cake; two pieces are two-thirds (2/3). The whole cake is three thirds: 3/3 = 1.

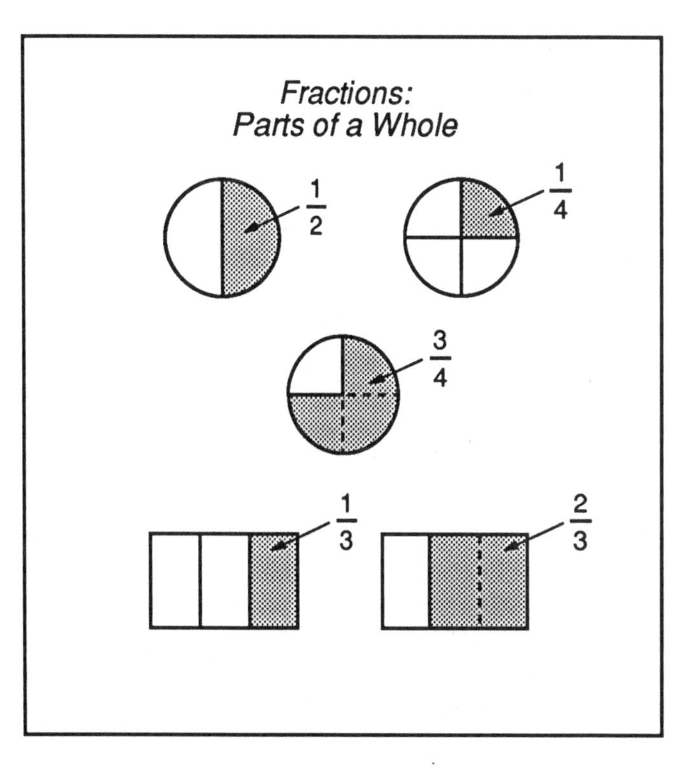

and may be written in different ways.

A fraction is a division—3/4 means 3 divided by 4. There are two other ways of writing divisions, but they're used for a division that is meant to be performed. A fraction like 3/4, though an "unfinished division," is meant to remain a fraction. You can divide 4 into 3, as we'll see later. But 3/4 can be added to other numbers, subtracted, etc., and used as a perfectly good number in its own right.

If you divide a pie into 4 equal parts and take 2 of the parts, you'd have 2/4 of the pie. Isn't this the same as 1/2? Yes: a fraction is equivalent to many others—just divide the whole into more parts and take more parts. This amounts to multiplying top and bottom by the same number, which doesn't change the fraction. For example, 1/2 = 2/4, 3/6, 4/8, ...; 1/3 = 2/6, 3/9, 4/12, ...; 3/4 = 6/8, 9/12, 12/16, ... Or you can divide top and bottom by the same number to reduce the fraction to "lowest terms": divide top and bottom by 3 to change 3/6 to 1/2.

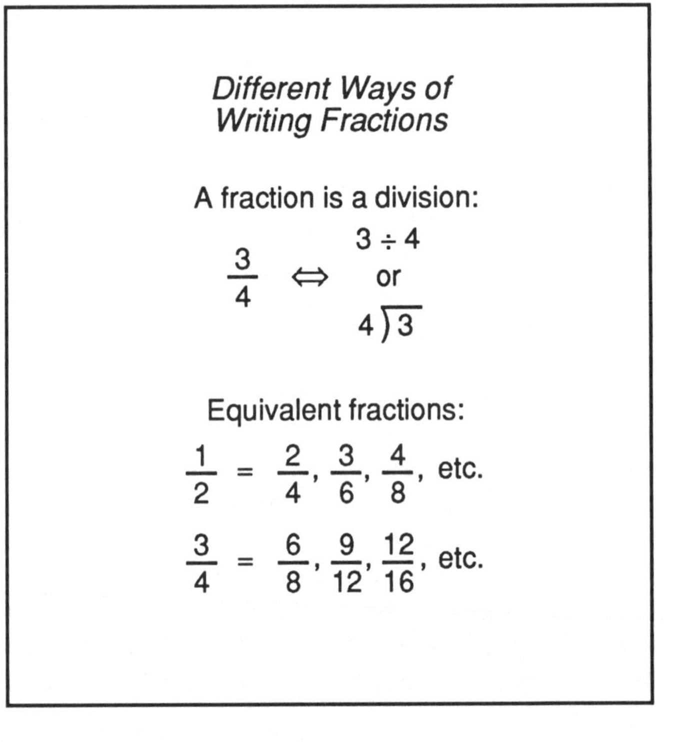

Adding or subtracting fractions is easy, IF—...

Adding and subtracting fractions is easy enough if the fractions have the same bottom (denominator). To add one-fifth (1/5) and two-fifths (2/5) is like adding one apple and two apples: the result is three-fifths (3/5), as it would be three apples. Same for subtraction: taking one-third (1/3) from two-thirds (2/3) leaves one-third (1/3). The reason is that the denominator is a kind of "bunch size"—it indicates the size of the things referred to. In whole numbers, this is shown by the position of a digit in the number (place value); in fractions, it is shown by the denominator. When the denominators are different, you first need to find a common measure—a common denominator. We'll see how in a moment.

Since 3/5 and 4/5 have the same denominator, you can add them: 3/5 + 4/5 = 7/5. You can change this fraction, which is greater than 1, to a "mixed number" (integer and fraction): 7/5 = 5/5 + 2/5 = 1 2/5.

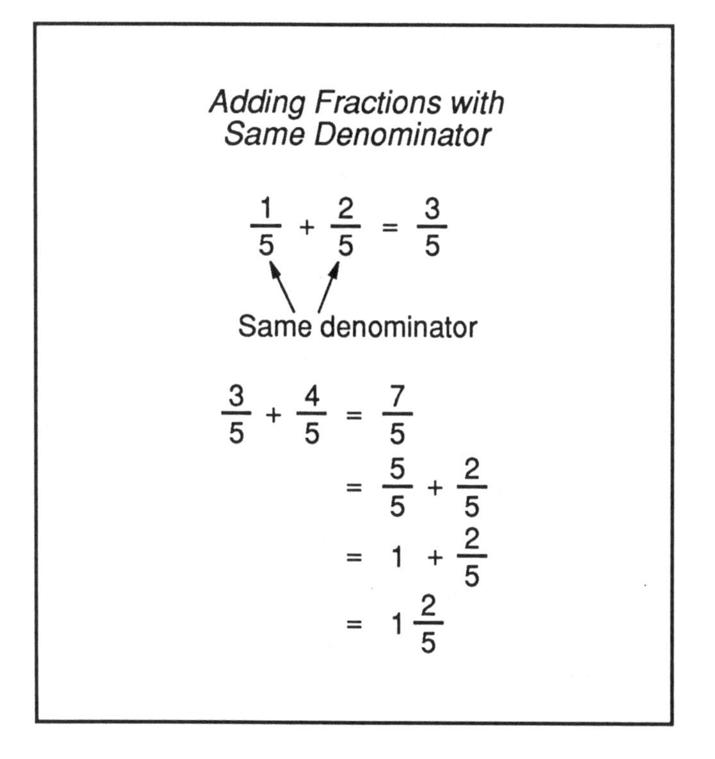

if they have a common denominator—...

To add 1/2 and 1/3, you need a common denominator (CD)—a bottom number that's the same for both fractions. This is used to express both fractions in terms of the same "bunch size." One way of finding a CD for 1/2 and 1/3 is to look at the two sequences 1/2 = 2/4, 3/6, 4/8, ..., and 1/3 = 2/6, 3/9, 4/12, ... Since 3/6 and 2/6 have the same denominator, 6, they can be added: 3/6 + 2/6 = 5/6.

To find a CD for two fractions, the easiest thing to do is multiply the two denominators. Sometimes this will give you a larger number than you need, but it always works. To add 2/3 and 3/4, multiply 3×4 to get 12 as the CD. Then to express 2/3 with 12 in the bottom, multiply the top by the same number, 4, you're multiplying the bottom by: $2 \times 4 = 8$, so 2/3 becomes 8/12. For 3/4, since you're multiplying the bottom by 3 to get 12, you must also multiply the top by 3: $3 \times 3 = 9$, so 3/4 = 9/12. Now you can add: 8/12 + 9/12 = 17/12, or 1 5/12.

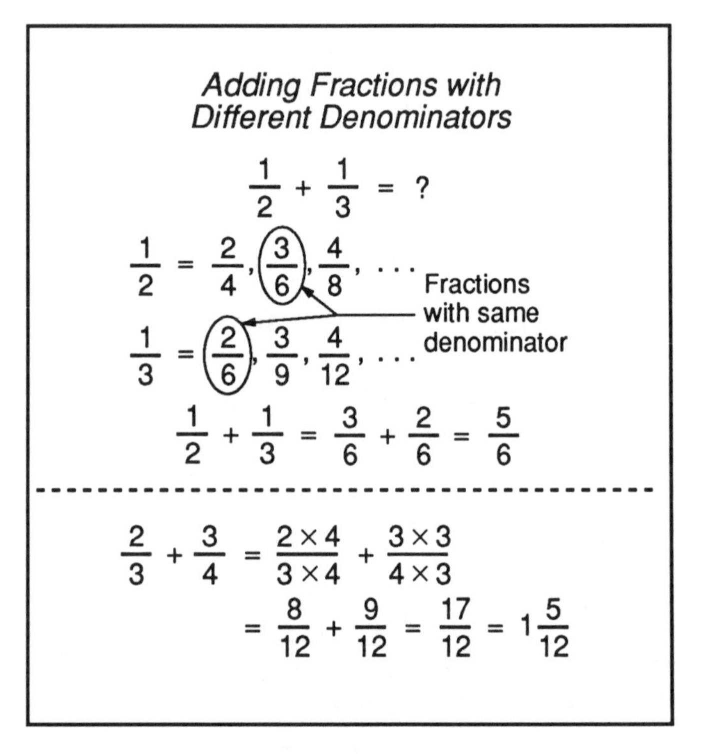

but multiplying by fractions is always easy.

When you multiply by a fraction, you're doing two things: multiplying by one number (the numerator) and dividing by another number (the denominator). This can be done in any order. So, for example, to multiply by 1/2, just divide by 2. To multiply (3/4) × 8 (or find "3/4 of 8"), multiply 3 × 8, which is 24, and divide by 4; the result is 6. Or divide first if you like: 8 ÷ 4 = 2, and 3 × 2 = 6. When both numbers are fractions, you're doing the same two things: multiplying and dividing. To multiply (2/3) × (4/5), first multiply 4/5 by 2, then divide by 3. This amounts to multiplying the tops together and multiplying the bottoms together: that is, (2 × 4)/(3 × 5), or 8/15. That's what makes multiplying by fractions so easy.

In multiplying fractions you can cancel terms that appear in the numerator and denominator: with (1/3) × (3/4), the 3's cancel since you're both multiplying and dividing by 3. Answer: 1/4.

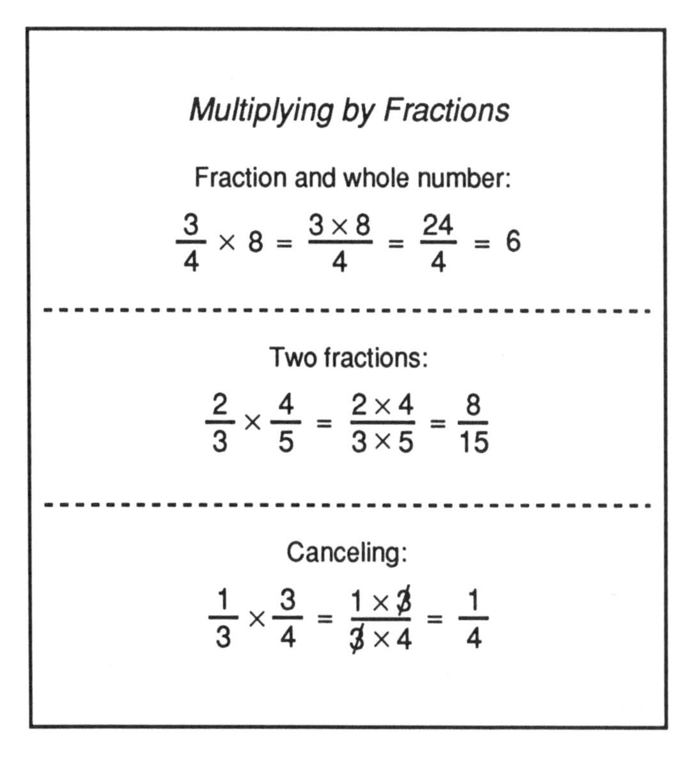

Dividing by a fraction is a little tricky,...

How do you divide by a fraction? For example, how do you divide 1 by 1/2? This is asking how many times 1/2 goes into 1—how many halves of a pie are in the whole pie. The answer is 2.

You know the answer here, but how do you get there? That is, if you see a problem like 1 ÷ (1/2), what do you do next? The usual way is to form a fraction, 1/(1/2), where the numerator is 1 and the denominator is itself a fraction, 1/2. Now 1/(1/2) is a rather awkward-looking fraction, since it has a fraction as the denominator. We need to get rid of that fraction in the denominator. How? Since the denominator is 1/2, the easiest way is to multiply it by 2, to give us 1 in the denominator. But we also have to multiply the numerator by 2 to compensate. (Multiplying top and bottom of a fraction by the same number does not change the value of the fraction.) We then have simply 2/1, or just 2—the answer we expected all along.

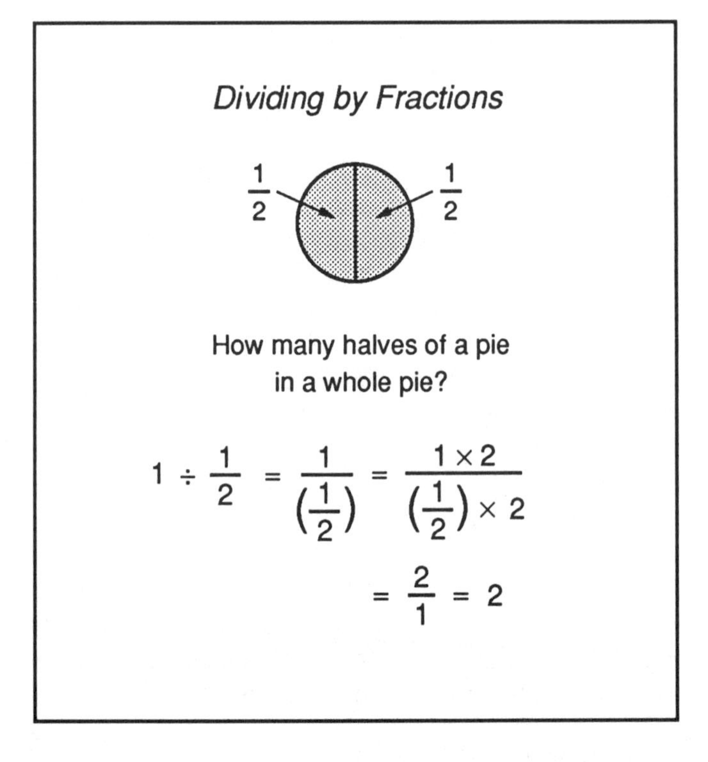

but a simple step...

Let's try another one. How about 1 ÷ (1/4)? The answer must be 4, since there are 4 quarters of a pie in the whole pie. Again, just form the fraction 1/(1/4), and get rid of that 1/4 in the denominator by multiplying top and bottom by 4. This gives us 4/1 or simply 4. Suppose you had two pies. How many quarters of a pie in two pies—that is, what is 2/(1/4)? You get the answer, 8, by multiplying top and bottom by 4.

When you divide 1 by a number, you get the *reciprocal* of the number. So the reciprocal of 1/2 is 2, and the reciprocal of 1/4 is 4 (also, the reciprocal of 2 is 1/2 and the reciprocal of 4 is 1/4). What we've found is that to divide by a fraction, you multiply by its reciprocal: 2/(1/4) = 2 × (4/1) = 8. (In fact, to divide by *any* number, you can multiply by its reciprocal: to divide by 10, multiply by its reciprocal, 1/10.) So the simple step of using the reciprocal changes division to multiplication.

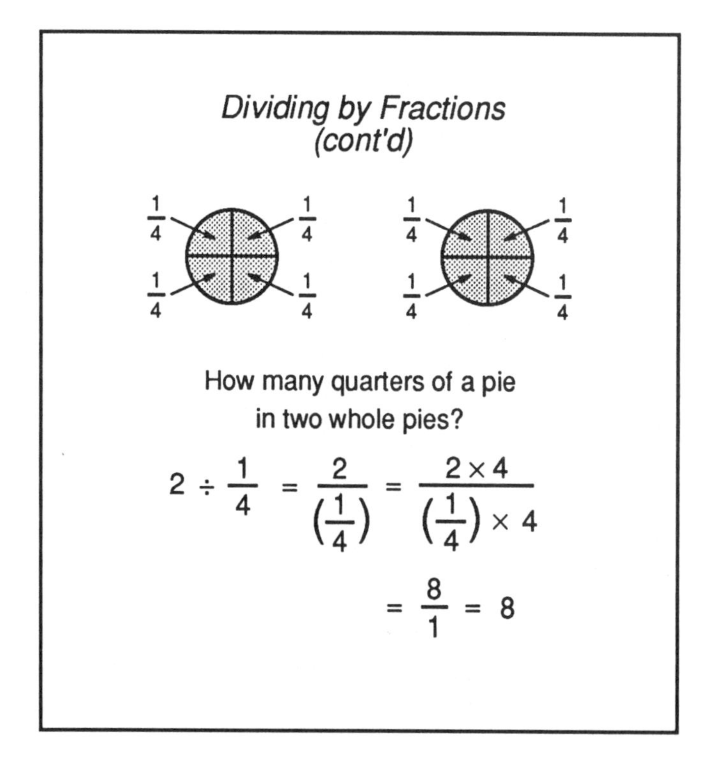

changes division into multiplication.

Here's a more complicated example: 3/4 ÷ 2/3. This forms an ugly looking fraction, (3/4)/(2/3). To get rid of that 2/3 in the denominator, let's multiply top and bottom by 3/2. This gives us 1 in the denominator—good. The numerator is (3/4) × (3/2), which we can multiply out to get (3 × 3)/(4 × 2), or 9/8. You can think of it this way: If you have a piece of string 3/4 of a foot (9 inches) long and another piece 2/3 of a foot (8 inches) long, the first piece is 9/8 as long as the second.

What we've done, again, is to multiply top and bottom of the ugly fraction by the reciprocal of the fraction in the denominator. Here the fraction in the denominator is 2/3, and its reciprocal is 3/2. We know it's the reciprocal because (2/3) × (3/2) = 1; that is, 3/2 is 1 divided by 2/3. To find the reciprocal of any fraction, invert top and bottom. And if a division problem has a fraction in the denominator, just invert the fraction and multiply.

Dividing by Fractions (cont'd)

$$\frac{3}{4} \div \frac{2}{3} = ?$$

$$\frac{3}{4} \div \frac{2}{3} = \frac{\left(\frac{3}{4}\right)}{\left(\frac{2}{3}\right)} = \frac{\left(\frac{3}{4}\right) \times \left(\frac{3}{2}\right)}{\left(\frac{2}{3}\right) \times \left(\frac{3}{2}\right)}$$

$$= \frac{\left(\frac{3}{4}\right) \times \left(\frac{3}{2}\right)}{1}$$

$$= \frac{(3 \times 3)}{(4 \times 2)} = \frac{9}{8} \text{ or } 1\frac{1}{8}$$

Decimals are like dollars and cents,...

In talking about money we like to talk about parts of a dollar and not just whole dollars. A dot—the decimal point—lets us do this. The dot between 6 and 4 in $356.42 shows where whole dollars end and parts of a dollar begin. The digits 3, 5, 6, 4 and 2, of course, have different values, from hundreds of dollars (3) down to hundredths of a dollar or pennies (2).

When we're not talking about money we don't use a $ sign, of course, but the same meaning applies to the digits: three hundreds, five tens, six ones, four tenths, and two hundredths. In other words, the decimal point lets us keep right on going to smaller numbers after reaching ones. Since the value of each digit is only one-tenth that of the digit to its left, to the right of the decimal point the first digit (in this case 4) has a value only one-tenth that of a one, or 1/10; the next digit (2) has a value only one-tenth of that, or 1/100, and so on.

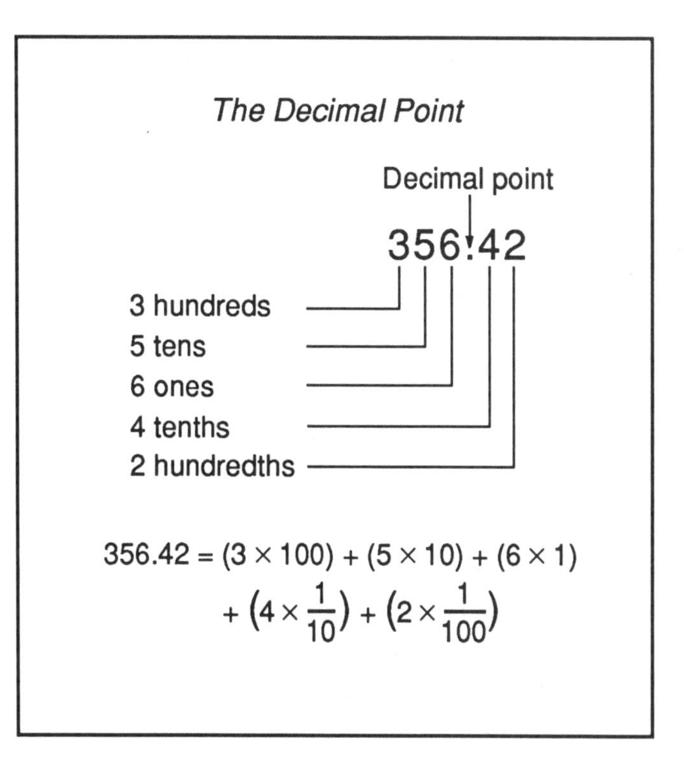

and make handling of fractions easy.

The decimal point has not caused such words as "half-dollar" and "quarter" to disappear from the language, but it does let us express everything on a common basis. If you use fractions in adding a half-dollar, a quarter, and a dime, you'd have to work a little to find you had 17/20 of a dollar. But if you think of them in the usual way as $0.50, $0.25, and $0.10, you can just add them to get $0.85. (The decimal point can be easily missed, so use a 0 to its left, as shown here.) Notice that 0.85, which is 85/100, is 17/20 when reduced to lowest terms (by dividing top and bottom by 5).

For amounts of money, two decimal places to the right are sufficient. But for other uses smaller and smaller numbers can be formed by continuing to write digits to the right of the decimal point, each digit having a value one-tenth as large as the one to its left. And, of course, larger numbers can be written to the left of the decimal point.

Decimal Fractions:
A Common Basis

	FRACTION	DECIMAL
Half-dollar	$\$\frac{1}{2}$	$0.50
Quarter	$\frac{1}{4}$	0.25
Dime	$+\frac{1}{10}$	+ 0.10
	$\$\frac{17}{20}$	$0.85

$$0.85 = \frac{85}{100} = \frac{17}{20} \text{ (reduced to lowest terms)}$$

Multiplying by 10 is no problem,...

As mentioned earlier (p. 98), to multiply a whole number—say, 274—by 10, just add a zero to the right: 2,740. To multiply by 100, add two zeros to the right: 27,400. And so on. What if the number has a decimal point?

Instead of the whole number 274, let's say the number is 274.596. Its meaning, like that of 274, is based on powers of 10 (such as 10, 100, 1,000, etc.). Only now we must include as powers of 10 such numbers as 1/10 (one-tenth), 1/100 (one-hundredth), and so on. So the meaning of 274.596 is 2 hundreds, 7 tens, 4 ones, 5 tenths, 9 hundredths, and 6 thousandths. If you multiply 274.596 by 10, it becomes—term by term—2 thousands, 7 hundreds, 4 tens, 5 ones, 9 tenths, and 6 hundredths, or 2,745.96. So to multiply a decimal number by 10, move the decimal point one place to the right. To multiply by 100 (or 10^2), move the decimal point two places to the right, and so on.

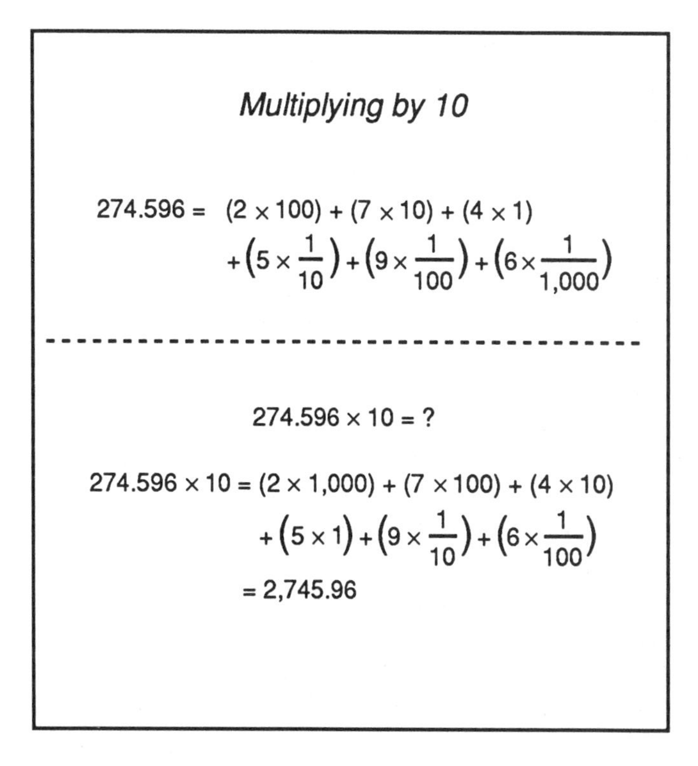

nor is dividing by 10;...

With decimal numbers, dividing by 10 is as easy as multiplying by 10. With whole numbers you can't divide by 10 unless you allow for a remainder, since a number such as 274 is not evenly divisible by 10. That is, if you divide 274 by 10 you'll get a quotient of 27 and a remainder of 4. But since there is a decimal point understood to be at the right of 274, to divide by 10 you can just move the decimal point one place to the left; that is, $274 \div 10 = 27.4$. And, of course, $27.4 \times 10 = 274$.

Taking a more complicated number like 274.596, you can divide by 10 just as easily. The number, which is 2 hundreds, 7 tens, 4 ones, 5 tenths, 9 hundredths, and 6 thousandths, becomes 2 tens, 7 ones, 4 tenths, 5 hundredths, 9 thousandths, and 6 ten-thousandths, or 27.4596. So to divide by 10, move the decimal point one place to the left; to divide by 100, move the decimal point two places to the left; and so on.

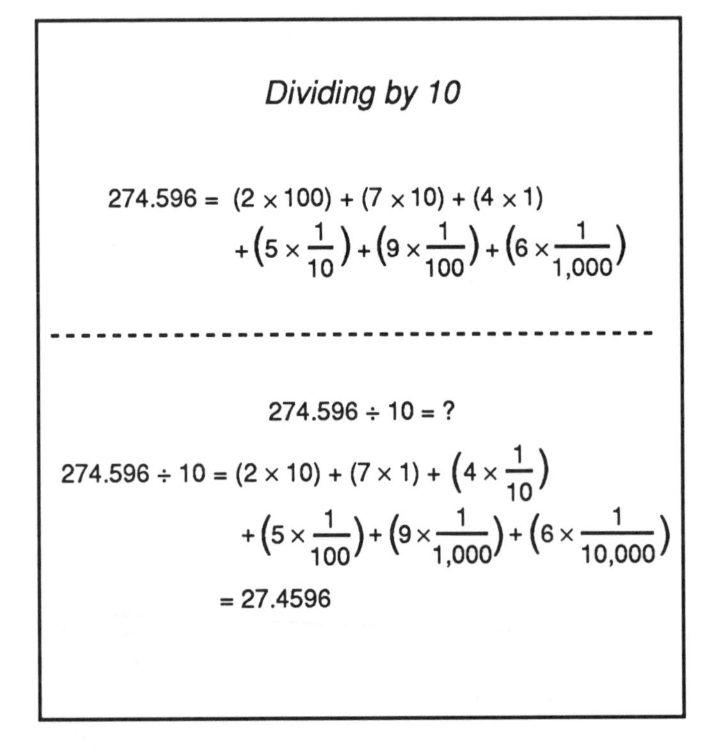

always, the decimal point is the key.

Adding and subtracting decimal numbers is the same as with whole numbers (for a whole number, the decimal point is at the right: 5 = 5. = 5.0). With a problem like 0.2 + 0.04, the decimal points must be lined up. Since 0.2 = 0.20 (that is, 2/10 = 20/100), 0.2 + 0.04 = 0.20 + 0.04 = 0.24. As for multiplying, recall that with whole numbers, the number of zeros in a product depends on the number in each factor. For the same reason, the number of places in a decimal product depends on the number in each factor, so 0.2 × 0.04 = (2/10) × (4/100) = 8/1,000, which is 0.008. There are as many places to the right of the decimal point in the product as there are in both factors together.

How about division: 0.2 ÷ 0.04? Multiply top and bottom of 0.2/0.04 by 100, to get rid of the decimal point in 0.04. The result is 20/4, or 5. The decimal point is moved two places to the right for 0.2, to make it 20, since there are two places in 0.04.

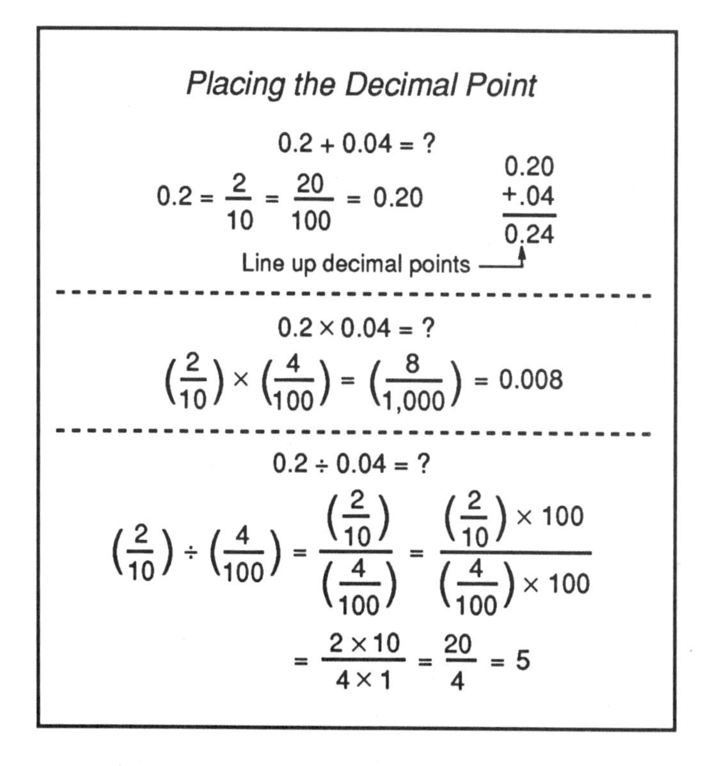

A fraction can be converted to a decimal,...

As mentioned earlier, fractions can be written in many ways: 1/2 = 2/4, 3/6, 4/8, 5/10, etc. But 5/10 is a decimal, 0.5. You can arrive at this by multiplying top and bottom of 1/2 by 5. You can also look at it a somewhat different way: Fractions, as mentioned earlier, are a form of "unfinished division." Using decimals, you can go on to finish the division: just divide the top by the bottom. Since 1/2 indicates 1 divided by 2, multiply top and bottom by 10. Then 1/2 becomes (10/2) × (1/10), or 5 × (1/10), or 0.5. It's as easy as dividing 10 by 2, but just remember that decimal point.

Similarly, 1/4 becomes 0.25. It's like dividing 100 by 4, but instead of getting 25, remember the decimal point and move it over two places. The fraction 3/4 becomes 0.75, which you can get either by multiplying 1/4 (which is 0.25) by 3, or by dividing 3 by 4—that is, divide 300 by 4 and move the decimal point.

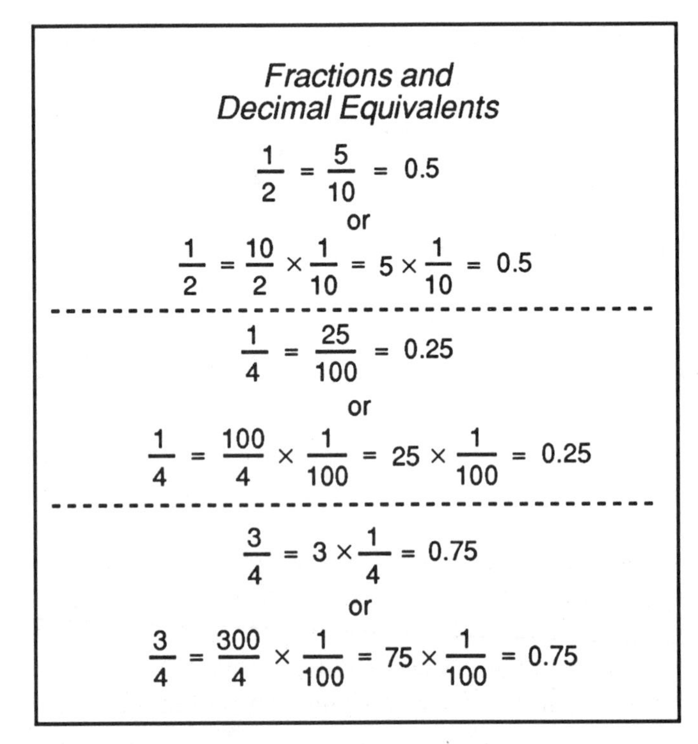

but the decimal may not terminate.

Some fractions don't convert to simple decimals. For example, 1/3 converts to 0.333..., with the 3's going on forever. The fraction 1/6 is 0.1666..., with the 6's going on forever after the initial 1. An interesting one is 1/7, which converts to a six-digit repeating decimal, 0.142857142857...; the six digits 142857 repeat forever. The fraction 1/9 converts to 0.111... To find fractions whose numerator is not 1, just multiply: 2/3 is $2 \times (1/3)$, or 0.666...; 5/9 is $5 \times (1/9)$, or 0.555...; and so on.

Any fraction, when converted, either terminates (as in 1/2 = 0.5) or has a repeating decimal. If the denominator is a multiple of 2 or 5 (which divide evenly into 10) the decimal terminates; otherwise it repeats. Decimals that don't terminate are "rounded off" to two or three places; for example, to three places, 1/3 is rounded off to 0.333, 2/3 to 0.667. If the following digit is less than 5, the third digit is left as is; if not, it's increased by 1.

Non-Terminating Decimals

$$\frac{1}{3} = ?$$

$$\frac{1}{3} = \frac{10}{3} \times \frac{1}{10} = 3.333... \times \frac{1}{10} = 0.333...$$

or

$$\frac{1}{3} = 0.333 \text{ (rounded off)}$$

$$\frac{1}{6} = 0.1666... = 0.167$$

$$\frac{1}{7} = 0.142857142857... = 0.143$$

$$\frac{1}{9} = 0.111... = 0.111$$

Fraction/decimal equivalents help you multiply.

Though you don't need to add fractions in your head very often, they're useful—particularly the decimal equivalents of fractions, which are handy for multiplying. Suppose you want to multiply 25 by 32. Realizing that 1/4 = 0.25, you see that 25 is 100/4, so you can divide 32 by 4 and multiply by 100. The answer you arrive at immediately: 800.

You may ask, "But how often do I multiply by 25?" Not very often. But each fraction has a decimal equivalent, and they can be used together with different powers of 10 to include more numbers; for example, 0.25 can become 2.5, 25, 250, etc. Some familiar fractions and their decimal equivalents (rounded to three places) are listed here. Only fractions whose numerator is 1 are listed; to find other fractions, multiply by the numerator: for example, 3/8 is $3 \times (1/8)$, or 3×0.125, or 0.375. And notice the relations between them: 1/12 is half of 1/6, 1/3 is three times 1/9, and so on.

Some Common Equivalents

$\frac{1}{2} = 0.5$

$\frac{1}{3} = 0.333$

$\frac{1}{4} = 0.25$

$\frac{1}{5} = 0.2$

$\frac{1}{6} = 0.167$

$\frac{1}{7} = 0.143$

$\frac{1}{8} = 0.125$

$\frac{1}{9} = 0.111$

$\frac{1}{10} = 0.1$

$\frac{1}{11} = 0.091$

$\frac{1}{12} = 0.083$

"Percent" means "per hundred,"...

Common fractions can have any number for a denominator: for 1/2 it's 2, for 2/3 it's 3, for 1/5 it's 5. In decimal fractions, we've seen, the denominator is always a power of 10. It's not shown explplictly but instead by the digit's position—0.4 is 4/10, 0.35 is 35/100, 0.008 is 8/1,000. A *percent* is a type of fraction in which the denominator is not stated but is always assumed to be 100. Thus 25 percent, or 25%, is 25/100; 5% is 5/100, and so on.

Some things have 100 parts—for example, a dollar has 100 cents. For such things, percent is obvious: 25 percent of a dollar is 25 cents; 5 percent is 5 cents, and so on. Suppose, however, there are not 100 parts—for example, a classroom may have 50 students. We can still speak of, say, 30 percent, as if the class had 100 students. Then 30% would mean 30 of the 100 students; but since the actual class has only half as many—50 students—30% would be 15 students.

Percent: Parts per Hundred

$$25\% = \frac{25}{100} \qquad 5\% = \frac{5}{100}$$

A class has 50 students.
How many is 30%?

$$30\% = \frac{30}{100} = \frac{15}{50}$$

= 15 students out of 50

and is easily changed to a fraction or decimal,...

A percent is thus already a fraction; it is a particular type of fraction in which the denominator is 100 but is not written out explicitly. So to change a percent to a fraction, just make the 100 explicit by writing it in the denominator, and reduce the fraction to lowest terms. A percent is also easily changed to decimal form; just divide by 100 by moving the decimal point two places to the left.

For example, 25% is 25/100. To change to a decimal, divide by 100 by moving the decimal point two places to the left: 0.25. To find the fraction, divide top and bottom of 25/100 by 25 to get 1/4. Another: 50% is 50/100—move the decimal point to get decimal 0.50 or 0.5 (since 5/10 = 50/100, 0.5 = 0.50), and divide top and bottom by 50, getting 1/2. Others: 5% = 5/100 = 0.05 = 1/20 (divide top and bottom by 5); 36% = 36/100 = 0.36 = 9/25 (divide top and bottom by 4); and 160% = 160/100 = 1.60 = 8/5 (divide top and bottom by 20).

Converting Percents

		DECIMAL	FRACTION
25%	$= \frac{25}{100} =$	0.25	$\frac{1}{4}$
50%	$= \frac{50}{100} =$	0.50	$\frac{1}{2}$
5%	$= \frac{5}{100} =$	0.05	$\frac{1}{20}$
36%	$= \frac{36}{100} =$	0.36	$\frac{9}{25}$
160%	$= \frac{160}{100} =$	1.60	$\frac{8}{5}$

even for non-terminating decimals.

Many fractions, as we've seen, when converted to decimals, produce a decimal that doesn't terminate: 1/3 = 0.333..., 1/6 = 0.1666..., 2/3 = 0.666..., and so on. How do we write these as percents?

Let's take 1/3. Since we want 100 in the denominator for a percent, let's multiply the decimal 0.333... by 100 and divide by 100 (multiplying and dividing by the same number doesn't change the value). We now have 100 in the denominator: (33.333...)/100. This rather awkward-looking fraction is actually 33.333... percent. But 33.333... can be written as 33 + 1/3, or 33 1/3. So the usual way of writing 1/3 as a percent is 33 1/3%. Other fractions that convert to repeating decimals are written in a similar way: 1/6 is (16.666...)/100, or 16 2/3%; 2/3 is (66.666...)/100, or 66 2/3%; 1/12 is (8.333...)/100 or 8 1/3%, and so on. Using a whole number and fraction combined (a mixed number) avoids the need for dots or round off.

Converting Percents (cont'd)

FRACTION	DECIMAL		PERCENT
$\frac{1}{3}$	= 0.333...	= $\frac{33.333...}{100}$	= $33\frac{1}{3}\%$
$\frac{1}{6}$	= 0.1666...	= $\frac{16.666...}{100}$	= $16\frac{2}{3}\%$
$\frac{2}{3}$	= 0.666...	= $\frac{66.666...}{100}$	= $66\frac{2}{3}\%$
$\frac{1}{12}$	= 0.08333...	= $\frac{8.333...}{100}$	= $8\frac{1}{3}\%$

Percent questions are of several types.

Percents sometimes cause trouble because it may not be clear whether to multiply or divide, or by what numbers. Percent questions tend to be of the following types:

- What is 25% of 32?
- 12 is what percent of 36?
- 7 is 20% of what?

These questions are all related and spring from the relation "A% of B is C," or $A \times B = C$. The first question asks about C, the second about A, and the third about B. To answer the first question, multiply 25%, or 1/4, by 32 (that is, divide 32 by 4) and you have 8. For the second question, you need to divide 12 by 36. Think of it as a fraction: 12/36 is 1/3, or 33 1/3%. The final question asks what number 7 is 20% of. That is, 7 is 20%, or 1/5, of some number. Just multiply 7×5 (or divide 7 by 1/5) and you have the number, 35.

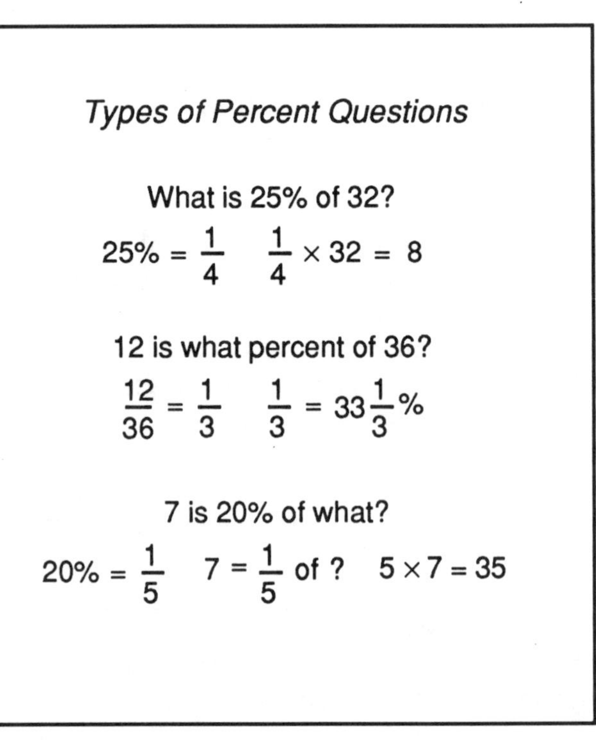

GLOSSARY

Many of the terms below have not been used in this book, since technical terms have been avoided where possible. But as they commonly appear elsewhere, it is worthwhile being familiar with them.

addend: one of a group of numbers to be added; in $4 + 5 = 9$, 4 and 5 are addends and 9 is the sum.

algebra: generalized arithmetic, in which letters are used to designate arbitrary numbers; for example, $a + b = b + a$ is the commutative law of addition, with a and b representing any two numbers.

Arabic numerals: the numerals we use in our number system—0,1,2,3,4,5,6,7,8,9; arrived in Europe from Arabia, probably having originated in India.

associative law of addition: a law stating that, for any numbers a, b, and c, $a + (b + c) = (a + b) + c$; for example, $4 + (2 + 7) = (4 + 2) + 7$, or $4 + 9 = 6 + 7$. In words: numbers can be grouped in any way when being added.

associative law of multiplication: a law stating that, for any numbers a, b, and c, $a \times (b \times c) = (a \times b) \times c$; for example, $5 \times (3 \times 8) = (5 \times 3) \times 8$, or $5 \times 24 = 15 \times 8$. In words: numbers can be grouped in any way when being multiplied.

base: the number of units in a digit's place that must be counted before producing a change of 1 in the next higher place; in our decimal system, the base is 10.

cancel: to divide the same number out of the numerator and denominator of a fraction, or the result of adding two numbers of opposite sign.

cardinal number: a number referring to the number

of things in a group without regard to order; for example, in speaking of "5 people," 5 is a cardinal number. Contrasted with ordinal number, which is concerned with order or sequence.

common denominator: a number that is a multiple of (or is divisible by) each denominator in a group of fractions; 20 is a common denominator for 1/4 and 1/5, since it is a multiple of both 4 and 5.

common fraction: a fraction whose numerator and denominator are both integers; for example, 2/3. Contrasted with a complex fraction, which has a fraction for the numerator or denominator, or both.

common divisor: a number that divides each of two or more other numbers; a common divisor of 6 and 15 is 3.

common multiple: a number that is a multiple of each of two or more other numbers; a common multiple of 6 and 4 is 12.

commutative law of addition: a law stating that, for any numbers a and b, $a + b = b + a$; for example, $8 + 5 = 5 + 8$. In words: order does not matter in adding.

commutative law of multiplication: a law stating that, for any numbers a and b, $a \times b = b \times a$; for example, $3 \times 7 = 7 \times 3$. In words: order does not matter in multiplying.

complex fraction: a fraction whose numerator or denominator is a fraction; for example, (1/3)/(4/5). Contrasted with a common fraction, whose numerator and denominator are both integers.

composite number: a number that can be factored into its components; an example is 6, which can be written as 2×3. Contrasted with a prime number, which has no factors.

decimal fraction: a fraction whose denominator is not written but is a power of 10 (such as 10, 100, 1,000, etc.); also called simply "decimal."

decimal number: any number containing a decimal point; for example, 3.14 or 0.007. Also, any number written in our standard decimal system.

decimal place: the position of a digit to the right of the decimal point; for example, in 3.14, 1 is in the first decimal place and 4 in the second.

decimal point: a dot that separates the units' place from the fractional part of a decimal number; in 3.14, the decimal point is between 3 and 1.

decimal system: our number system, which uses 10 as the base.

denominator: the number on the bottom (that is, below the line) of a fraction; the denominator of 3/5 is 5 and the numerator is 3.

difference: the result of subtracting one number from another; in $5 - 3 = 2$, 2 is the difference, 5 is the minuend, and 3 is the subtrahend.

digit: any of the numerals 0,1,2,3,4,5,6,7,8,9 used to write a number; in 516, the digits are 5, 1, and 6.

distributive law: a law stating that, for any three numbers a, b, and c, $a \times (b + c) = (a \times b) + (a \times c)$; for example, $3 \times (5 + 6) = (3 \times 5) + (3 \times 6)$, or $3 \times 11 = 15 + 18$.

dividend: the number that is being divided by another number; in $6 \div 3 = 2$, 6 is the dividend, 3 is the divisor, and 2 is the quotient.

divisor: the number by which another number is to be divided; in $21 \div 7 = 3$, 7 is the divisor, 21 is the dividend, and 3 is the quotient.

exponent: a small number written above and to the right of another number, indicating how many times the latter number is to appear as a factor; $4^2 = 4 \times 4$, $10^3 = 10 \times 10 \times 10$. An expression such as 4^2 is also referred to as "4 squared"; 10^3 is also "10 cubed," "the third power of 10," or "10 to the third power."

factor: used both as a verb and a noun, and applies only to whole numbers (or integers). As a verb: to factor a number is to break it down into factors; to factor 18, write it as 2×9 or $2 \times 3 \times 3$. As a noun: a factor of a number is a number which, when multiplied by another number, yields the first number as a product. Using the same example, 9 is a factor of 18 because it can be multiplied by 2 to give 18.

factorization theorem: a theorem stating that any integer can be expressed in one and only one way as a product of its prime factors; for example, $12 = 2 \times 2 \times 3$.

fraction: a division (or quotient) of two numbers, written with the dividend on top and the divisor on the bottom, separated by a line. The dividend is called the numerator and the divisor the denominator. In the fraction 2/3, 2 is the numerator and 3 the denominator.

greatest common divisor: the largest common divisor of two or more numbers; usually abbreviated as GCD. The GCD of 16 and 28 is 4.

identity: a statement of equality which is true for all values of the symbols appearing in the statement; for example, $a + b = b + a$.

improper fraction: a fraction in which the numerator is larger than, or equal to, the denominator; 5/4 is an improper fraction. Contrasted with proper fraction, in which the numerator is less than the denominator.

inequality: a statement that one number is less than or greater than another; for two numbers a and b, $a < b$ means that a is less than b, and $a > b$ means a is greater than b. For example, $2 < 5$ and $8 > 6$.

integer: any of the numbers 0, 1, 2, 3, ...; also called whole number. May also refer to negative integers, –1, –2, –3, ... Contrasted with fractions and decimals.

inverse: an operation that nullifies a previous operation; subtraction of a number reverses a prior addition of that same number, and division by a number reverses multiplication by that number.

irrational number: a number that cannot be expressed as an integer or a fraction; an example is $\sqrt{2}$ (the square root of 2). Also includes transcendental numbers such as $\pi = 3.14159...$ (where the numbers do not repeat), the ratio of the circumference to diameter of a circle.

least common multiple: the smallest number that is divisible by each of two or more other numbers; usually abbreviated as LCM. The LCM of 2, 3, and 4 is 12.

lowest terms: a fraction in which the common divisors of the numerator and denominator have been divided out; 1/2 is in lowest terms, but 5/10 is not, since top and bottom can be divided by 5.

minuend: the number from which another number is being subtracted; in $8 - 3 = 5$, 8 is the minuend, 3 is the subtrahend, and 5 is the difference.

mixed number: a number written as a combination of integer and fraction; for example, 3 5/8.

multiple: an integer that is a product of the stated number and another integer; 10 is a multiple of 2 (it is also a multiple of 5). A number is a multiple of any of its factors.

multiplicand: the number that is being multiplied by another number; in $7 \times 4 = 28$, 4 is the multiplicand, 7 is the multiplier, and 28 is the product.

multiplier: the number by which another number is being multiplied; in $3 \times 13 = 39$, 3 is the multiplier, 13 is the multiplicand, and 39 is the product.

numerals: the symbols used to write numbers; examples are our Arabic numerals (0,1,2,...) or Roman numerals (I, V, X, ...).

numerator: the number that is on the top (that is, above the line) of a fraction; the numerator of 3/5 is 3 and the denominator is 5.

ordinal number: a number that denotes order or sequence; in "February 15, 1992" 15 and 1992 are ordinal numbers. Contrasted with cardinal number, which is used to denote quantity.

partial product: an intermediate product produced in digit-by-digit multiplication; in multiplying 13×45, the results of multiplying 3×45 and 10×45 are partial products.

percent: parts per hundred, denoted by %; 17% of a number is 17/100 of it.

place value: the value a digit has because of its position or place in a number; the value of 7 in 27 is 7 units, in 74 is 7 tens, and in 735 is 7 hundreds.

power: the value of an exponent; 10 to the fourth power is 10^4, which is $10 \times 10 \times 10 \times 10$, or 10,000.

prime number: an integer that cannot be divided by any integers but itself and 1; examples are 2, 3, 5, 7, 11, and 13. Contrasted with composite numbers such as 4, 6, 8, 9, 10, and 12, which have divisors. All primes other than 2 are odd; 1 is not considered a prime.

product: the result of multiplying two or more numbers together; in $9 \times 5 = 45$, 45 is the product, 9 is the multiplier, and 5 is the multiplicand.

proper fraction: a fraction in which the numerator is less than the denominator; 3/4 is a proper fraction. Contrasted with an improper fraction, in which the numerator is larger than or equal to the denominator.

proportion: the equality of two ratios, written a:b = c:d or a/b = c/d, where a, b, c and d are numbers; for example, 3/5 = 6/10, or 3 is to 5 as 6 is to 10.

quotient: the result of dividing one number by another; in $24 \div 6 = 4$, 4 is the quotient, 24 is the dividend, and 6 is the divisor.

ratio: the relative sizes of two numbers, as given by a fraction or quotient, as a/b, where a and b are numbers; the ratio of 4 to 7 is 4/7.

rational number: a number that can be expressed as an integer or fraction; 5 and 1/2 are rational numbers. Contrasted with irrational numbers, such as $\sqrt{2}$, which cannot be so expressed.

reciprocal: the number formed by dividing 1 by a given number; the reciprocal of 2 is 1/2, and of 3/4 is 4/3. The product of a number and its reciprocal is 1.

remainder: the number left over when dividing one integer by another; $15 \div 6 = 2$ with remainder 3. Since $15 - (6 \times 2) = 3$, the remainder is the dividend minus the product of the divisor and quotient.

repeating decimal: a decimal in which all the digits after a certain point consist of a group that repeats indefinitely; 0.333..., 0.1666... , and 0.0909... are repeating decimals. Contrasted with non-terminating, non-repeating decimals, such as produced by irrational numbers like $\sqrt{2}$ and π. Any rational number can be expressed as a terminating or repeating decimal.

root: a number which, when raised to the proper power, produces the given number; 4 is the third or "cube" root of 64, since $4^3 = 4 \times 4 \times 4 = 64$.

rounding off: dropping decimals after a certain number of places. If the first digit dropped is less than 5, the preceding digit is left as is; if it's 5 or greater, the preceding digit is increased by 1. Thus 0.333... is rounded to 0.33 to two places; 0.1666... is rounded to 0.17 to two places and to 0.167 to three places.

square: the result of multiplying a number by itself; the square of 10 is 10×10, or 10^2, or 100.

square root: a number which, when multiplied by itself, produces the given number; the square root of 25 is 5, since $5^2 = 5 \times 5 = 25$.

subtrahend: a number that is being subtracted from another number; in $13 - 9 = 4$, 9 is the subtrahend, 13 is the minuend, and 4 is the difference.

sum: the result of adding two or more numbers; in $23 + 49 = 72$, 72 is the sum and 23 and 49 are addends.

terminating decimal: a decimal fraction whose digits end after a certain point; 0.125 is a terminating decimal. Contrasted with non-terminating decimals such as 0.666..., which continue indefinitely.

whole number: any of the numbers 0, 1, 2, 3, ...; also called integer. Contrasted with fractions and decimals.

INDEX

ABOUT THE AUTHOR

W. J. ("Jim") Howard is a writer who specializes in making technical information understandable to the general reader. Formerly a mathematician with the Rand Corporation and Planning Research Corporation, he has done consulting work for General Electric, TRW, Aerospace Corporation, Matrix Corporation, Serendipity Associates, Xerox Corporation, Xyzyx Information Corporation and others. He has produced a variety of books, many under contract. Some examples: *Vegetable Gardening; Kitchen and Bathroom Remodeling;* and *A Simple Manual on Queues* (Xyzyx Information Corporation—Dr. Kay Inaba); *Venereal Disease*, co-authored with Dr. Ruth Schlesinger (USC School of Medicine—Dr. Art Ulene); *Introduction to Interactive Accounting System* and *Turn Around Time: A Key Factor in Printing Productivity* (Xerox Corporation). His latest book (1991) is *Life's Beginnings: Our Life Before Birth Helps Us Trace Our Roots.*

The present book was typeset and illustrated by the author, with the help of an Apple Macintosh® computer and LaserWriter® printer, and Aldus PageMaker® and Adobe Illustrator™ software.